BRITISH RA 

4

EMU

RAIL

TWENTY-FOURTH EDITION
2011

The Complete Guide to all
Electric Multiple Units which operate on
the national railway network and the stock of
the major UK Light Rail & Metro systems

Robert Pritchard & Peter Fox

ISBN 978 1902 336 82 4

Printed in England by Information Press, Eynsham, Oxford.

CONTENTS

PROVISION OF INFORMATION

This book has been compiled with care to be as accurate as possible, but in some cases information is not officially available and the publisher cannot be held responsible for any errors or omissions. We would like to thank the companies and individuals which have been co-operative in supplying information to us. The authors of this series of books are always pleased to receive notification from readers of any inaccuracies readers may find in the series, to enhance future editions. Please send comments to:

Robert Pritchard, Platform 5 Publishing Ltd., 3 Wyvern House, Sark Road, Sheffield, S2 4HG, England.
e-mail: robert@platform5.com **Tel:** 0114 255 2625 **Fax:** 0114 255 2471

This book is updated to information received by 4 October 2010.

UPDATES

This book is updated to the Stock Changes given in **Today's Railways UK 107** (November 2010). Readers are therefore advised to update this book from the official Platform 5 Stock Changes published every month in **Today's Railways UK** magazine, starting with issue 108.

The Platform 5 magazine **Today's Railways UK** contains news and rolling stock information on the railways of Britain and Ireland and is published on the second Monday of every month. For further details of **Today's Railways UK**, please see the advertisement on the back cover of this book.

Front cover photograph: London Overground-liveried 378 228 arrives at Kensal Green with the 10.21 Watford Junction–London Euston on 10/07/10. **Jason Cross**

BRITAIN'S RAILWAY SYSTEM

INFRASTRUCTURE & OPERATION

Britain's national railway infrastructure is owned by a "not for dividend" company, Network Rail. Many stations and maintenance depots are leased to and operated by Train Operating Companies (TOCs), but some larger stations remain under Network Rail control. The only exception is the infrastructure on the Isle of Wight, which is nationally owned and is leased to South West Trains.

Trains are operated by TOCs over Network Rail, regulated by access agreements between the parties involved. In general, TOCs are responsible for the provision and maintenance of the locos, rolling stock and staff necessary for the direct operation of services, whilst NR is responsible for the provision and maintenance of the infrastructure and also for staff to regulate the operation of services.

DOMESTIC PASSENGER TRAIN OPERATORS

The large majority of passenger trains are operated by the TOCs on fixed term franchises. Franchise expiry dates are shown in the list of franchisees below:

Franchise	*Franchisee*	*Trading Name*
Chiltern Railways	Deutsche Bahn (until 31 December 2021)	Chiltern Railways
Cross-Country[1]	Deutsche Bahn (Arriva) (until 1 November 2013)	CrossCountry
East Midlands[2]	Stagecoach Holdings plc (until 11 November 2013)	East Midlands Trains
Greater Western[3]	First Group plc (until 1 April 2013)	First Great Western
Greater Anglia	National Express Group plc (until 14 October 2011)	National Express East Anglia
Integrated Kent[4]	GoVia Ltd. (Go-Ahead/Keolis) (until 3 March 2012)	Southeastern
InterCity East Coast[5]		East Coast
InterCity West Coast	Virgin Rail Group Ltd. (until 31 March 2012)	Virgin Trains
London Rail[6]	MTR/Deutsche Bahn (until 14 March 2014)	London Overground
LTS Rail[7]	National Express Group plc (until further notice)	c2c
Merseyrail Electrics[8]	Serco/NedRail (until 20 July 2028)	Merseyrail
Northern Rail	Serco/Abellio (until 15 September 2013)	Northern
ScotRail	First Group plc (until 8 November 2014)	ScotRail
South Central[9]	GoVia Ltd. (Go-Ahead/Keolis) (until 25 July 2015)	Southern

South Western[10]	Stagecoach Holdings plc (until 4 February 2014)	South West Trains
Thameslink/Great Northern[11]	First Group plc (until 1 April 2012)	First Capital Connect
Trans-Pennine Express[12]	First Group/Keolis (until 1 February 2012)	TransPennine Express
Wales & Borders	Deutsche Bahn (Arriva) (until 6 December 2018)	Arriva Trains Wales
West Midlands[13]	GoVia Ltd. (Go-Ahead/Keolis) (until 19 September 2013)	London Midland

Notes:

[1] Awarded for six years to 2013 with an extension for a further two years and five months to 1 April 2016 if performance targets are met.

[2] Awarded for six years to 2013 with an extension for a further one year and five months to 1 April 2015 if performance targets are met.

[3] Awarded for seven years to 2013 with an extension for a further three years to 1 April 2016 if performance targets are met.

[4] The Integrated Kent franchise started on 1 April 2006 for an initial period of six years to 2012, with an extension for a further two years to 1 April 2014 if performance targets are met.

[5] Currently run on an interim basis by DfT management company Directly Operated Railways (trading as East Coast) following financial difficulties experienced by National Express Group.

[6] The London Rail Concession is different from all other rail franchises, as fares and service levels are set by Transport for London instead of the DfT. Incorporates the North and West London lines, the Gospel Oak–Barking line and Euston–Watford local services.

[7] Tendering for the new LTS Rail franchise, to be called Essex Thameside, has been delayed whilst the Government conducts a consultation exercise on the future of rail franchising policy (the franchise has been due to finish on 25 May 2011). The Greater Anglia franchise has also been given a short extension for the same reason.

[8] Now under control of Merseytravel PTE instead of the DfT. Franchise due to be reviewed after seven years (in July 2010) and then every five years to fit in with the Merseyside Local Transport Plan.

[9] Awarded for five years and ten months to 2015 with a possible extension for a further two years to 25 July 2017.

[10] Awarded for seven years to 2014 with an extension for a further three years to 4 February 2017 if performance targets are met.

[11] Awarded for six years to 2012 with an extension for up to a further three years to 1 April 2015 if performance targets are met.

[12] Awarded for eight years to 2012 with an extension a further five years to 1 February 2017 if performance targets are met.

[13] Awarded for six years to 2013 with an extension for a further two years to 19 September 2015 if performance targets are met.

All new franchises officially start at 02.00 on the first day. Because of this the finishing date of an old franchise and the start date of its successor are the same.

Where termination dates are dependent on performance targets being met, the earliest possible termination date is given. However, with Merseyrail the termination date is based on the maximum franchise length.

The following operators run non-franchised services only:

Operator	*Trading Name*	*Route*
BAA	Heathrow Express	London Paddington–Heathrow Airport
First Hull Trains	First Hull Trains	London King's Cross–Hull
Grand Central	Grand Central	London King's Cross–Sunderland/ Bradford Interchange
North Yorkshire Moors Railway Enterprises	North Yorkshire Moors Railway	Pickering–Grosmont–Whitby/ Battersby
West Coast Railway Company	West Coast Railway Company	Birmingham–Stratford-upon-Avon Fort William–Mallaig* York–Leeds–York–Scarborough* Machynlleth–Porthmadog/Pwllheli*
Wrexham, Shropshire & Marylebone Railway	Wrexham & Shropshire	London Marylebone–Wrexham General

* Special summer-dated services only.

INTERNATIONAL PASSENGER OPERATIONS

Eurostar (UK) operates passenger services between the UK and mainland Europe, jointly with the national operators of France (SNCF) and Belgium (SNCB/NMBS). Eurostar (UK) is a subsidiary of London & Continental Railways, which is jointly owned by National Express Group and British Airways.

In addition, a service for the conveyance of accompanied road vehicles through the Channel Tunnel is provided by the tunnel operating company, Eurotunnel.

FREIGHT TRAIN OPERATIONS

The following operators operate freight services or empty passenger stock workings under "Open Access" arrangements:

Colas Rail
DB Schenker Rail (UK)
Direct Rail Services (DRS)
Europorte2 (Eurotunnel)
Freightliner
GB Railfreight (owned by Eurotunnel)
West Coast Railway Company

INTRODUCTION

EMU CLASSES

Principal details and dimensions are quoted for each class in metric and/o imperial units as considered appropriate bearing in mind common UK usage

All dimensions and weights are quoted for vehicles in an "as new" conditior with all necessary supplies on board. Dimensions are quoted in the orde length x overall width. All lengths quoted are over buffers or couplers as appropriate. Where two lengths are quoted, the first refers to outer vehicles in a set and the second to inner vehicles.

Bogie Types are quoted in the format motored/non-motored (e.g BP20/BT13 denotes BP20 motored bogies and BT non-motored bogies).

Unless noted to the contrary, all vehicles listed have bar couplers at non driving ends.

Vehicles ordered under the auspices of BR were allocated a Lot (batch) numbe when ordered and these are quoted in class headings and sub-headings Vehicles ordered since 1995 have no Lot Numbers, but the manufacturer anc location that they were built is given.

NUMERICAL LISTINGS

25 kV AC 50 Hz overhead Electric Multiple Units (EMUs) and dual voltage EMUs are listed in numerical order of set numbers. Individual "loose" vehicles are listed in numerical order after vehicles formed into fixed formations.

750 V DC third rail EMUs are listed in numerical order of class number, ther in numerical order of set number. Some of these use the former Southerr Region four-digit set numbers. These are derived from theoretical six digi set numbers which are the four-digit set number prefixed by the first twc numbers of the class.

Where sets or vehicles have been renumbered in recent years, forme numbering detail is shown alongside current detail. Each entry is laid out as in the following example:

Set No.	*Detail*	*Livery*	*Owner*	*Operator*	*Allocation*	*Formation*
365 505	*	**FU**	E	*FC*	HE	65898 72249 72248 65939

Detail Differences. Only detail differences which currently affect the areas and types of train which vehicles may work are shown. All other detai differences are specifically excluded. Where such differences occur within a class or part class, these are shown alongside the individual set or vehicle number. Meaning of abbreviations are detailed in individual class headings

Set Formations. Set formations shown are those normally maintained Readers should note some set formations might be temporarily varied from time to time to suit maintenance and/or operational requirements. Vehicle shown as "Spare" are not formed in any regular set formation.

Codes. Codes are used to denote the livery, owner, operator and depot of each unit. Details of these will be found in section 7 of this book. Where a unit or spare car is off-lease, the operator column will be left blank.

Names. Only names carried with official sanction are listed. As far as possible names are shown in UPPER/lower case characters as actually shown on the name carried on the vehicle(s). Unless otherwise shown, complete units are regarded as named rather than just the individual car(s) which carry the name.

GENERAL INFORMATION

CLASSIFICATION AND NUMBERING

25 kV AC 50 Hz overhead and "Versatile" EMUs are classified in series 300–399.

750 V DC third rail EMUs are classified in the series 400–599.

Service units are classified in the series 900–949.

EMU individual cars are numbered in the series 61000–78999, except for vehicles used on the Isle of Wight – which are numbered in a separate series, and for the Class 378s, 380s and 395s, which will take up new 38xxx and 39xxx series'.

Any vehicle constructed or converted to replace another vehicle following accident damage and carrying the same number as the original vehicle is denoted by the suffix II in this publication.

OPERATING CODES

These codes are used by train operating company staff to describe the various different types of vehicles and normally appear on data panels on the inner (i.e. non driving) ends of vehicles.

A "B" prefix indicates a battery vehicle.
A "P" prefix indicates a trailer vehicle on which is mounted the pantograph, instead of the default case where the pantograph is mounted on a motor vehicle.

The first part of the code describes whether or not the car has a motor or a driving cab as follows:

DM	Driving motor	M	Motor
DT	Driving trailer	T	Trailer

The next letter is a "B" for cars with a brake compartment.
This is followed by the saloon details:

F	First	S	Standard
C	Composite		

The next letter denotes the style of accommodation as follows:

O	Open	K	Side compartment with lavatory

so Semi-open (part compartments, part open). All other vehicles are assumed to consist solely of open saloons.

Finally vehicles with a buffet are suffixed RB or RMB for a miniature buffet.

Where two vehicles of the same type are formed within the same unit, the above codes may be suffixed by (A) and (B) to differentiate between the vehicles

A composite is a vehicle containing both first and standard class accommodation, whilst a brake vehicle is a vehicle containing separate specific accommodation for the conductor.

Special Note: Where vehicles have been declassified, the correct operating code which describes the actual vehicle layout is quoted in this publication.

The following codes are used to denote special types of vehicle:

DMLF Driving Motor Lounge First
DMLV Driving Motor Luggage Van
MBRBS Motor buffet standard with luggage space and guard's compartment.
TFH Trailer First with Handbrake

BUILD DETAILS

Lot Numbers

Vehicles ordered under the auspices of BR were allocated a Lot (batch) number when ordered and these are quoted in class headings and sub-headings.

ACCOMMODATION

The information given in class headings and sub-headings is in the form F/S nT (or TD) nW. For example 12/54 1T 1W denotes 12 first class and 54 standard class seats, one toilet and one space for a wheelchair. A number in brackets (i.e. (2)) denotes tip-up seats (in addition to the fixed seats). Tip-up seats in vestibules do not count. The seating layout of open saloons is shown as 2+1, 2+2 or 3+2 as the case may be. Where units have first class accommodation as well as standard and the layout is different for each class then these are shown separately prefixed by "1:" and "2:". Compartments are three seats a side in first class and mostly four a side in standard class in EMUs. TD denotes a toilet suitable for use by a disabled person.

ABBREVIATIONS

The following standard abbreviations are used in class headings and also throughout this publication:

AC Alternating Current.
BR British Railways.
BSI Bergische Stahl Industrie.
DC Direct Current.
EMU Electric Multiple Unit.
Hz Hertz.
kN kilonewtons.
km/h kilometres per hour.
kW kilowatts.
LT London Transport.
LUL London Underground Limited.
m. metres.
m.p.h. miles per hour.
SR BR Southern Region.
t. tonnes.
V volts.

1. 25 kV AC 50 Hz OVERHEAD & DUAL VOLTAGE UNITS

Note: Except where otherwise stated, all units in this section operate on 25 kV AC 50 Hz overhead only.

CLASS 313 BREL YORK

Inner suburban units.

Formation: DMSO–PTSO–BDMSO.
Systems: 25 kV AC overhead/750 V DC third rail.
Construction: Steel underframe, aluminium alloy body and roof.
Traction Motors: Four GEC G310AZ of 82.125 kW.
Wheel Arrangement: Bo-Bo + 2-2 + Bo-Bo.
Braking: Disc & rheostatic.
Dimensions: 20.33/20.18 x 2.82 m.
Bogies: BX1.
Couplers: Tightlock.
Gangways: Within unit + end doors.
Control System: Camshaft.
Doors: Sliding.
Maximum Speed: 75 m.p.h.
Seating Layout: 313/0: Refurbished with high back seats (3+2 facing layout). 313/1 Low back seats, mainly 2+2 facing. 313/2s being refurbished for Southern with 2+2 high-back seating.
Multiple Working: Within class.

DMSO. Lot No. 30879 1976–1977. –/74 (313/1 –/66). 36.0 t.
PTSO. Lot No. 30880 1976–1977. –/83 (313/1 –/70 1W). 31.0 t.
BDMSO. Lot No. 30885 1976–1977. –/74 (313/1 –/66). 37.5 t.

Class 313/0. Standard Design.

313 018	**FU**	E	*FC*	HE	62546	71230	62610
313 024	**FU**	E	*FC*	HE	62552	71236	62616
313 025	**FU**	E	*FC*	HE	62553	71237	62617
313 026	**FU**	E	*FC*	HE	62554	71238	62618
313 027	**FU**	E	*FC*	HE	62555	71239	62619
313 028	**FU**	E	*FC*	HE	62556	71240	62620
313 029	**FU**	E	*FC*	HE	62557	71241	62621
313 030	**FU**	E	*FC*	HE	62558	71242	62622
313 031	**FU**	E	*FC*	HE	62559	71243	62623
313 032	**FU**	E	*FC*	HE	62560	71244	62643
313 033	**FU**	E	*FC*	HE	62561	71245	62625
313 035	**FU**	E	*FC*	HE	62563	71247	62627
313 036	**FU**	E	*FC*	HE	62564	71248	62628
313 037	**FU**	E	*FC*	HE	62565	71249	62629
313 038	**FU**	E	*FC*	HE	62566	71250	62630
313 039	**FU**	E	*FC*	HE	62567	71251	62631
313 040	**FU**	E	*FC*	HE	62568	71252	62632
313 041	**FU**	E	*FC*	HE	62569	71253	62633
313 042	**FU**	E	*FC*	HE	62570	71254	62634
313 043	**FU**	E	*FC*	HE	62571	71255	62635

313 044	**FU**	E	*FC*	HE	62572	71256	62636
313 045	**FU**	E	*FC*	HE	62573	71257	62637
313 046	**FU**	E	*FC*	HE	62574	71258	62638
313 047	**FU**	E	*FC*	HE	62575	71259	62639
313 048	**FU**	E	*FC*	HE	62576	71260	62640
313 049	**FU**	E	*FC*	HE	62577	71261	62641
313 050	**FU**	E	*FC*	HE	62578	71262	62649
313 051	**FU**	E	*FC*	HE	62579	71263	62624
313 052	**FU**	E	*FC*	HE	62580	71264	62644
313 053	**FU**	E	*FC*	HE	62581	71265	62645
313 054	**FU**	E	*FC*	HE	62582	71266	62646
313 055	**FU**	E	*FC*	HE	62583	71267	62647
313 056	**FU**	E	*FC*	HE	62584	71268	62648
313 057	**FU**	E	*FC*	HE	62585	71269	62642
313 058	**FU**	E	*FC*	HE	62586	71270	62650
313 059	**FU**	E	*FC*	HE	62587	71271	62651
313 060	**FU**	E	*FC*	HE	62588	71272	62652
313 061	**FU**	E	*FC*	HE	62589	71273	62653
313 062	**FU**	E	*FC*	HE	62590	71274	62654
313 063	**FU**	E	*FC*	HE	62591	71275	62655
313 064	**FU**	E	*FC*	HE	62592	71276	62656

Name (carried on PTSO): 313 054 Captain William Leefe Robinson V.C.

Class 313/1. Former London Overground units. At Brighton for Southern crew training (to be refurbished and renumbered 313/2s).

313 101	**SL**	E	BI	62529	71213	62593
313 108	**SL**	E	BI	62536	71220	62600

Class 313/1. Former London Overground units. 313 122/123/134 to transfer to First Capital Connect.

313 121	**SL**	E	ZN	62549	71233	62613
313 122	**SL**	E	HE	62550	71234	62614
313 123	**SL**	E	ZN	62551	71235	62615
313 134	**SL**	E	HE	62562	71246	62626

Name (carried on PTSO): 313 134 The Hackney Empire

Class 313/2. Southern units. 19 units are being refurbished for Southern for Brighton Coastway services with pantographs removed (some units in traffic unrefurbished).

Details for refurbished units (r):

DMSO. Lot No. 30879 1976–1977. –/64. . t.
PTSO. Lot No. 30880 1976–1977. –/68. . t.
BDMSO. Lot No. 30885 1976–1977. –/64. . t.

313 201	(313 101)								
313 202	(313 102)		**SN**	E	*SN*	BI	62530	71214	62594
313 203	(313 103)		**SN**	E	*SN*	BI	62531	71215	62595
313 204	(313 104)		**SN**	E	*SN*	BI	62532	71216	62596
313 205	(313 105)		**SN**	E	*SN*	BI	62533	71217	62597
313 206	(313 106)	r	**SN**	E	*SN*	BI	62534	71218	62598

313 207	(313 107)		**SN**	E	*SN*	BI	62535	71219	62599
313 208	(313 108)								
313 209	(313 109)	r	**SN**	E	*SN*	BI	62537	71221	62601
313 210	(313 110)	r	**SN**	E	*SN*	BI	62538	71222	62602
313 211	(313 111)	r	**SN**	E	*SN*	BI	62539	71223	62603
313 212	(313 112)		**SN**	E	*SN*	BI	62540	71224	62604
313 213	(313 113)	r	**SN**	E	*SN*	BI	62541	71225	62605
313 214	(313 114)	r	**SN**	E	*SN*	BI	62542	71226	62606
313 215	(313 115)		**SN**	E	*SN*	BI	62543	71227	62607
313 216	(313 116)	r	**SN**	E	*SN*	BI	62544	71228	62608
313 217	(313 117)	r	**SN**	E	*SN*	BI	62545	71229	62609
313 219	(313 119)	r	**SN**	E	*SN*	BI	62547	71231	62611
313 220	(313 120)		**SN**	E	*SN*	BI	62548	71232	62612

CLASS 314 BREL YORK

Inner suburban units.

Formation: DMSO–PTSO–DMSO.
Construction: Steel underframe, aluminium alloy body and roof.
Traction Motors: Four GEC G310AZ (* Brush TM61-53) of 82.125 kW.
Wheel Arrangement: Bo-Bo + 2-2 + Bo-Bo.
Braking: Disc & rheostatic.
Dimensions: 20.33/20.18 x 2.82 m.
Bogies: BX1.
Couplers: Tightlock.
Gangways: Within unit + end doors.
Control System: Thyristor.
Doors: Sliding.
Maximum Speed: 70 m.p.h.
Seating Layout: 3+2 low-back facing.
Multiple Working: Within class and with Class 315.

DMSO. Lot No. 30912 1979. –/68. 34.5 t.
64588ll. DMSO. Lot No. 30908 1978–1980. Rebuilt Railcare Glasgow 1996 from Class 507 No. 64426. The original 64588 has been scrapped. –/74. 34.5 t.
PTSO. Lot No. 30913 1979. –/76. 33.0 t.

314 201	*	**SC**	A	*SR*	GW	64583	71450	64584	
314 202	*	**SC**	A	*SR*	GW	64585	71451	64586	
314 203	*	**SC**	A	*SR*	GW	64587	71452	64588ll	European Union
314 204	*	**SC**	A	*SR*	GW	64589	71453	64590	
314 205	*	**SC**	A	*SR*	GW	64591	71454	64592	
314 206	*	**SC**	A	*SR*	GW	64593	71455	64594	
314 207		**SC**	A	*SR*	GW	64595	71456	64596	
314 208		**SC**	A	*SR*	GW	64597	71457	64598	
314 209		**SC**	A	*SR*	GW	64599	71458	64600	
314 210		**SC**	A	*SR*	GW	64601	71459	64602	
314 211		**SC**	A	*SR*	GW	64603	71460	64604	
314 212		**SC**	A	*SR*	GW	64605	71461	64606	
314 213		**SC**	A	*SR*	GW	64607	71462	64608	
314 214		**SC**	A	*SR*	GW	64609	71463	64610	
314 215		**SC**	A	*SR*	GW	64611	71464	64612	
314 216		**SC**	A	*SR*	GW	64613	71465	64614	

CLASS 315 BREL YORK

Inner suburban units.

Formation: DMSO–TSO–PTSO–DMSO.
Construction: Steel underframe, aluminium alloy body and roof.
Traction Motors: Four Brush TM61-53 (* GEC G310AZ) of 82.125 kW.
Wheel Arrangement: Bo-Bo + 2-2 + 2-2 + Bo-Bo.
Braking: Disc & rheostatic.
Dimensions: 20.33/20.18 x 2.82 m.
Bogies: BX1.
Couplers: Tightlock.
Gangways: Within unit + end doors.
Control System: Thyristor.
Doors: Sliding.
Maximum Speed: 75 m.p.h.
Seating Layout: 3+2 low-back facing.
Multiple Working: Within class and with Class 314.

DMSO. Lot No. 30902 1980–1981. –/74. 35.0 t.
TSO. Lot No. 30904 1980–1981. –/86. 25.5 t.
PTSO. Lot No. 30903 1980–1981. –/84. 32.0 t.

315 801	**1**	E	*EA*	IL	64461	71281	71389	64462
315 802	**1**	E	*EA*	IL	64463	71282	71390	64464
315 803	**1**	E	*EA*	IL	64465	71283	71391	64466
315 804	**1**	E	*EA*	IL	64467	71284	71392	64468
315 805	**1**	E	*EA*	IL	64469	71285	71393	64470
315 806	**1**	E	*EA*	IL	64471	71286	71394	64472
315 807	**1**	E	*EA*	IL	64473	71287	71395	64474
315 808	**1**	E	*EA*	IL	64475	71288	71396	64476
315 809	**1**	E	*EA*	IL	64477	71289	71397	64478
315 810	**1**	E	*EA*	IL	64479	71290	71398	64480
315 811	**1**	E	*EA*	IL	64481	71291	71399	64482
315 812	**1**	E	*EA*	IL	64483	71292	71400	64484
315 813	**1**	E	*EA*	IL	64485	71293	71401	64486
315 814	**1**	E	*EA*	IL	64487	71294	71402	64488
315 815	**1**	E	*EA*	IL	64489	71295	71403	64490
315 816	**1**	E	*EA*	IL	64491	71296	71404	64492
315 817	**1**	E	*EA*	IL	64493	71297	71405	64494
315 818	**1**	E	*EA*	IL	64495	71298	71406	64496
315 819	**1**	E	*EA*	IL	64497	71299	71407	64498
315 820	**1**	E	*EA*	IL	64499	71300	71408	64500
315 821	**1**	E	*EA*	IL	64501	71301	71409	64502
315 822	**1**	E	*EA*	IL	64503	71302	71410	64504
315 823	**1**	E	*EA*	IL	64505	71303	71411	64506
315 824	**1**	E	*EA*	IL	64507	71304	71412	64508
315 825	**1**	E	*EA*	IL	64509	71305	71413	64510
315 826	**1**	E	*EA*	IL	64511	71306	71414	64512
315 827	**1**	E	*EA*	IL	64513	71307	71415	64514
315 828	**1**	E	*EA*	IL	64515	71308	71416	64516
315 829	**1**	E	*EA*	IL	64517	71309	71417	64518
315 830	**1**	E	*EA*	IL	64519	71310	71418	64520
315 831	**1**	E	*EA*	IL	64521	71311	71419	64522
315 832	**1**	E	*EA*	IL	64523	71312	71420	64524

315 833	1	E	*EA*	IL	64525	71313	71421	64526
315 834	1	E	*EA*	IL	64527	71314	71422	64528
315 835	1	E	*EA*	IL	64529	71315	71423	64530
315 836	1	E	*EA*	IL	64531	71316	71424	64532
315 837	1	E	*EA*	IL	64533	71317	71425	64534
315 838	1	E	*EA*	IL	64535	71318	71426	64536
315 839	1	E	*EA*	IL	64537	71319	71427	64538
315 840	1	E	*EA*	IL	64539	71320	71428	64540
315 841	1	E	*EA*	IL	64541	71321	71429	64542
315 842	* 1	E	*EA*	IL	64543	71322	71430	64544
315 843	* 1	E	*EA*	IL	64545	71323	71431	64546
315 844	* 1	E	*EA*	IL	64547	71324	71432	64548
315 845	* 1	E	*EA*	IL	64549	71325	71433	64550
315 846	* 1	E	*EA*	IL	64551	71326	71434	64552
315 847	* 1	E	*EA*	IL	64553	71327	71435	64554
315 848	* 1	E	*EA*	IL	64555	71328	71436	64556
315 849	* 1	E	*EA*	IL	64557	71329	71437	64558
315 850	* 1	E	*EA*	IL	64559	71330	71438	64560
315 851	* 1	E	*EA*	IL	64561	71331	71439	64562
315 852	* 1	E	*EA*	IL	64563	71332	71440	64564
315 853	* 1	E	*EA*	IL	64565	71333	71441	64566
315 854	* 1	E	*EA*	IL	64567	71334	71442	64568
315 855	* 1	E	*EA*	IL	64569	71335	71443	64570
315 856	* 1	E	*EA*	IL	64571	71336	71444	64572
315 857	* 1	E	*EA*	IL	64573	71337	71445	64574
315 858	* 1	E	*EA*	IL	64575	71338	71446	64576
315 859	* 1	E	*EA*	IL	64577	71339	71447	64578
315 860	* 1	E	*EA*	IL	64579	71340	71448	64580
315 861	* 1	E	*EA*	IL	64581	71341	71449	64582

Names (carried on DMSO):

315 817 Transport for London
315 829 London Borough of Havering Celebrating 40 years
315 845 Herbie Woodward
315 857 Stratford Connections

CLASS 317 — BREL YORK/DERBY

Outer suburban units.

Formation: Various, see sub-class headings.
Construction: Steel.
Traction Motors: Four GEC G315BZ of 247.5 kW.
Wheel Arrangement: 2-2 + Bo-Bo + 2-2 + 2-2.
Braking: Disc.
Bogies: BP20 (MSO), BT13 (others).
Gangways: Throughout
Doors: Sliding.
Dimensions: 20.13/20.18 x 2.82 m.
Couplers: Tightlock.
Control System: Thyristor.
Maximum Speed: 100 m.p.h.
Seating Layout: Various, see sub-class headings.
Multiple Working: Within class & with Classes 318, 319, 320, 321, 322 and 323.

Class 317/1. Pressure ventilated.

Formation: DTSO–MSO–TCO–DTSO.
Seating Layout: 1: 2+2 facing, 2: 3+2 facing.

DTSO(A) Lot No. 30955 York 1981–1982. –/74. 29.5 t.
MSO. Lot No. 30958 York 1981–1982. –/79. 49.0 t.
TCO. Lot No. 30957 Derby 1981–1982. 22/46 2T. 29.0 t.
DTSO(B) Lot No. 30956 York 1981–1982. –/71. 29.5 t.

317 337	**FU**	A	*FC*	HE	77036	62671	71613	77084
317 338	**FU**	A	*FC*	HE	77037	62698	71614	77085
317 339	**FU**	A	*FC*	HE	77038	62699	71615	77086
317 340	**FU**	A	*FC*	HE	77039	62700	71616	77087
317 341	**FU**	A	*FC*	HE	77040	62701	71617	77088
317 342	**FU**	A	*FC*	HE	77041	62702	71618	77089
317 343	**FU**	A	*FC*	HE	77042	62703	71619	77090
317 344	**FU**	A	*FC*	HE	77029	62690	71620	77091
317 345	**FU**	A	*FC*	HE	77044	62705	71621	77092
317 346	**FU**	A	*FC*	HE	77045	62706	71622	77093
317 347	**FU**	A	*FC*	HE	77046	62707	71623	77094
317 348	**FU**	A	*FC*	HE	77047	62708	71624	77095

Names (carried on TCO):

317 345 Driver John Webb | 317 348 Richard A Jenner

Class 317/5. Pressure ventilated. Units renumbered from Class 317/1 in 2005 for West Anglia Metro services. Refurbished with new upholstery and Passenger Information Systems. Details as Class 317/1.

317 501	**NX**	A	*EA*	IL	77024	62661	71577	77048
317 502	**NX**	A	*EA*	IL	77001	62662	71578	77049
317 503	**NX**	A	*EA*	IL	77002	62663	71579	77050
317 504	**NX**	A	*EA*	IL	77003	62664	71580	77051
317 505	**NX**	A	*EA*	IL	77004	62665	71581	77052
317 506	**NX**	A	*EA*	IL	77005	62666	71582	77053
317 507	**NX**	A	*EA*	IL	77006	62667	71583	77054
317 508	**NX**	A	*EA*	IL	77010	62697	71587	77058
317 509	**NX**	A	*EA*	IL	77011	62672	71588	77059
317 510	**NX**	A	*EA*	IL	77012	62673	71589	77060
317 511	**1**	A	*EA*	IL	77014	62675	71591	77062
317 512	**NC**	A	*EA*	IL	77015	62676	71592	77063
317 513	**NX**	A	*EA*	IL	77016	62677	71593	77064
317 514	**NX**	A	*EA*	IL	77017	62678	71594	77065
317 515	**NX**	A	*EA*	IL	77019	62680	71596	77067

Name (carried on TCO):

317 507 University of Cambridge 800 Years 1209–2009

Class 317/6. Convection heating. Units converted from Class 317/2 by Railcare Wolverton 1998–99 with new seating layouts.

Formation: DTSO–MSO–TSO–DTCO.
Seating Layout: 2+2 facing.

77200–77219. DTSO. Lot No. 30994 York 1985–1986. –/64. 29.5 t.
77280–77283. DTSO. Lot No. 31007 York 1987. –/64. 29.5 t.
62846–62865. MSO. Lot No. 30996 York 1985–1986. –/70. 49.0 t.
62886–62889. MSO. Lot No. 31009 York 1987. –/70. 49.0 t.
71734–71753. TSO. Lot No. 30997 York 1985–1986. –/62 2T. 29.0 t.
71762–71765. TSO. Lot No. 31010 York 1987. –/62 2T. 29.0 t.
77220–77239. DTCO. Lot No. 30995 York 1985–1986. 24/48. 29.5 t.
77284–77287. DTCO. Lot No. 31008 York 1987. 24/48. 29.5 t.

317 649	**WN**	A	*EA*	IL	77200	62846	71734	77220	
317 650	**WN**	A	*EA*	IL	77201	62847	71735	77221	
317 651	**WN**	A	*EA*	IL	77202	62848	71736	77222	
317 652	**1**	A	*EA*	IL	77203	62849	71739	77223	
317 653	**1**	A	*EA*	IL	77204	62850	71738	77224	
317 654	**1**	A	*EA*	IL	77205	62851	71737	77225	Richard Wells
317 655	**1**	A	*EA*	IL	77206	62852	71740	77226	
317 656	**1**	A	*EA*	IL	77207	62853	71742	77227	
317 657	**1**	A	*EA*	IL	77208	62854	71741	77228	
317 658	**1**	A	*EA*	IL	77209	62855	71743	77229	
317 659	**1**	A	*EA*	IL	77210	62856	71744	77230	
317 660	**1**	A	*EA*	IL	77211	62857	71745	77231	
317 661	**1**	A	*EA*	IL	77212	62858	71746	77232	
317 662	**1**	A	*EA*	IL	77213	62859	71747	77233	
317 663	**1**	A	*EA*	IL	77214	62860	71748	77234	
317 664	**1**	A	*EA*	IL	77215	62861	71749	77235	
317 665	**1**	A	*EA*	IL	77216	62862	71750	77236	
317 666	**1**	A	*EA*	IL	77217	62863	71752	77237	
317 667	**1**	A	*EA*	IL	77218	62864	71751	77238	
317 668	**1**	A	*EA*	IL	77219	62865	71753	77239	
317 669	**1**	A	*EA*	IL	77280	62886	71762	77284	
317 670	**1**	A	*EA*	IL	77281	62887	71763	77285	
317 671	**1**	A	*EA*	IL	77282	62888	71764	77286	
317 672	**1**	A	*EA*	IL	77283	62889	71765	77287	

Class 317/7. Units converted from Class 317/1 by Railcare Wolverton 2000 for Stansted Express services between London Liverpool Street and Stansted. Air conditioning. Fitted with luggage stacks.

Formation: DTSO–MSO–TSO–DTCO.
Seating Layout: 1: 2+1 facing, 2: 2+2 facing.

DTSO Lot No. 30955 York 1981–1982. –/52 + catering point. 31.4 t.
MSO. Lot No. 30958 York 1981–1982. –/62. 51.3 t.
TSO. Lot No. 30957 Derby 1981–1982. –/42 (5) 1W 1T 1TD. 30.2 t.
DTCO Lot No. 30956 York 1981–1982. 22/16 + catering point. 31.6 t.

317 708	**NX**	A	*EA*	IL	77007	62668	71584	77055
317 709	**NX**	A	*EA*	IL	77008	62669	71585	77056

317 710	**NX**	A	*EA*	IL	77009	62670	71586	77057	
317 714	**NX**	A	*EA*	IL	77013	62674	71590	77061	
317 719	**NX**	A	*EA*	IL	77018	62679	71595	77066	
317 722	**NX**	A	*EA*	IL	77021	62682	71598	77069	
317 723	**NX**	A	*EA*	IL	77022	62683	71599	77070	
317 729	**NX**	A	*EA*	IL	77028	62689	71605	77076	
317 732	**NX**	A	*EA*	IL	77031	62692	71608	77079	

Names (carried on DTCO):

317 709 Len Camp | 317 723 The Tottenham Flyer

Class 317/8. Pressure Ventilated. Units refurbished and renumbered from Class 317/1 in 2005–2006 at Wabtec, Doncaster for use on Stansted Express services. Fitted with luggage stacks.

Formation: DTSO–MSO–TCO–DTSO.
Seating Layout: 1: 2+2 facing, 2: 3+2 facing.

DTSO(A) Lot No. 30955 York 1981–1982. –/66. 29.5 t.
MSO. Lot No. 30958 York 1981–1982. –/71. 49.0 t.
TCO. Lot No. 30957 Derby 1981–1982. 20/42 2T. 29.0 t.
DTSO(B) Lot No. 30956 York 1981–1982. –/66. 29.5 t.

317 881	**NX**	A	*EA*	IL	77020	62681	71597	77068	
317 882	**NC**	A	*EA*	IL	77023	62684	71600	77071	
317 883	**NC**	A	*EA*	IL	77000	62685	71601	77072	
317 884	**NC**	A	*EA*	IL	77025	62686	71602	77073	
317 885	**NC**	A	*EA*	IL	77026	62687	71603	77074	
317 886	**NC**	A	*EA*	IL	77027	62688	71604	77075	
317 887	**NX**	A	*EA*	IL	77043	62704	71606	77077	
317 888	**NX**	A	*EA*	IL	77030	62691	71607	77078	
317 889	**NX**	A	*EA*	IL	77032	62693	71609	77080	
317 890	**NX**	A	*EA*	IL	77033	62694	71610	77081	
317 891	**NX**	A	*EA*	IL	77034	62695	71611	77082	
317 892	**NX**	A	*EA*	IL	77035	62696	71612	77083	Ilford Depot

CLASS 318 BREL YORK

Outer suburban units.

Formation: DTSO–MSO–DTSO.
Construction: Steel.
Traction Motors: Four Brush TM 2141 of 268 kW.
Wheel Arrangement: 2-2 + Bo-Bo + 2-2.
Braking: Disc.
Bogies: BP20 (MSO), BT13 (others).
Gangways: Within unit.
Doors: Sliding.
Dimensions: 19.83/19.92 x 2.82 m.
Couplers: Tightlock.
Control System: Thyristor.
Maximum Speed: 90 m.p.h.
Seating Layout: 3+2 facing.
Multiple Working: Within class & with Classes 317, 319, 320, 321, 322 and 323.

77240–77259. DTSO. Lot No. 30999 1985–1986. –/64 1T. 30.0 t.
77288. DTSO. Lot No. 31020 1987. –/64 1T. 30.0 t.

▲ Class 313s started to enter service with Southern on "Coastway" services from Brighton in 2010 (initially unrefurbished). On 24/05/10 313 202 arrives at Lewes with the 10.10 Brighton–Seaford. **Robert Pritchard**

▼ 314 203 "European Union" leaves Glasgow Central with the 16.15 Cathcart Circle service on 02/09/10. This unit is unique as its centre car is a former Class 507 vehicle. **Robert Pritchard**

▲ National Express-liveried 317 729 and 317 714 arrive at Stansted Airport with the 09.40 from London Liverpool Street on 09/08/10. New Class 379s start to work these services from spring 2011. **Robert Pritchard**

▼ Strathclyde PTE-liveried 318 262 approaches Lanark with the 16.23 from Anderston on 11/05/09. **Robin Ralston**

▲ First Capital Connect-liveried 319 375 (coupled to Southern-liveried 319 008) arrives at Brighton with the 16.10 from Bedford on 24/05/10. **Robert Pritchard**

▼ London Midland-liveried 321 412 and 321 411 approach Leighton Buzzard with the 18.05 Euston–Northampton on 24/05/10. **John Turner**

▲ Northern-liveried 323 224 passes Longport with the 18.57 Stoke-on-Trent–Manchester Piccadilly on 24/05/10. **Cliff Beeton**

▼ Royal Mail-liveried 325 013 passes Wandelmill, on the West Coast Main Line between Carstairs and Beattock, with 1M03 17.42 Shieldmuir–Warrington RMT mail on 14/04/10. **Robin Ralston**

▲ Heathrow Express units 332 014 and 332 011 are seen at Heathrow Airport Terminal 5 with the 10.12 to London Paddington on 08/06/10. **Robert Pritchard**

▼ Northern/West Yorkshire PTE-liveried 333 001 arrives at Guiseley with the 14.10 Ilkley–Leeds on 21/04/10. **Chris Wilson**

▲ Strathclyde PTE-liveried 334 001 is seen near Dalry with the 18.15 Glasgow Central–Ardrossan Town on 18/04/10. **Robin Ralston**

▼ London Midland-liveried 350 116/256 pass Roade with the 07.39 Northampton–London Euston on 04/06/10. **Dave Gommersall**

▲ National Express (white)-liveried 357 039 arrives at Barking with the 12.10 London Fenchurch Street–Shoeburyness on 24/07/10. **Robert Pritchard**

▼ Heathrow Connect-liveried 360 202 arrives at Hayes & Harlington with a service from Heathrow Airport Terminal 4 to London Paddington on 29/04/09. **Dave Gommersall**

▲ Southeastern's 375 307 leads a 7-car formation near Hither Green with the 12.10 London Charing Cross–Dover Priory/Canterbury West on 21/04/10.
Alex Dasi-Sutton

▼ Inner suburban unit 376 006 is seen near Petts Wood with the 14.20 London Cannon Street–Orpington on 21/05/10.
Alex Dasi-Sutton

62866–62885. MSO. Lot No. 30998 1985–1986. –/77. 50.9 t.
62890. MSO. Lot No. 31019 1987. –/77. 50.9 t.
77260–77279. DTSO. Lot No. 31000 1985–1986. –/72. 29.6 t.
77289. DTSO. Lot No. 31021 1987. –/72. 29.6 t.

318 250	**SC**	E	*SR*	GW	77240	62866	77260	
318 251	**SC**	E	*SR*	GW	77241	62867	77261	
318 252	**SC**	E	*SR*	GW	77242	62868	77262	
318 253	**SC**	E	*SR*	GW	77243	62869	77263	
318 254	**SC**	E	*SR*	GW	77244	62870	77264	
318 255	**SC**	E	*SR*	GW	77245	62871	77265	
318 256	**SC**	E	*SR*	GW	77246	62872	77266	
318 257	**SC**	E	*SR*	GW	77247	62873	77267	
318 258	**SC**	E	*SR*	GW	77248	62874	77268	
318 259	**SC**	E	*SR*	GW	77249	62875	77269	Citizens' Network
318 260	**SC**	E	*SR*	GW	77250	62876	77270	
318 261	**SC**	E	*SR*	GW	77251	62877	77271	
318 262	**SC**	E	*SR*	GW	77252	62878	77272	
318 263	**SC**	E	*SR*	GW	77253	62879	77273	
318 264	**SC**	E	*SR*	GW	77254	62880	77274	
318 265	**SC**	E	*SR*	GW	77255	62881	77275	
318 266	**SC**	E	*SR*	GW	77256	62882	77276	STRATHCLYDER
318 267	**SC**	E	*SR*	GW	77257	62883	77277	
318 268	**SC**	E	*SR*	GW	77258	62884	77278	
318 269	**SC**	E	*SR*	GW	77259	62885	77279	
318 270	**SC**	E	*SR*	GW	77288	62890	77289	

CLASS 319 BREL YORK

Express and outer suburban units.

Formation: Various, see sub-class headings.
Systems: 25 kV AC overhead/750 V DC third rail.
Construction: Steel.
Traction Motors: Four GEC G315BZ of 268 kW.
Wheel Arrangement: 2-2 + Bo-Bo + 2-2 + 2-2.
Braking: Disc.
Bogies: P7-4 (MSO), T3-7 (others).
Gangways: Within unit + end doors.
Doors: Sliding.
Dimensions: 20.17/20.16 x 2.82 m.
Couplers: Tightlock.
Control System: GTO chopper.
Maximum Speed: 100 m.p.h.
Seating Layout: Various, see sub-class headings.
Multiple Working: Within class & with Classes 317, 318, 320, 321, 322 and 323.

Class 319/0. DTSO–MSO–TSO–DTSO.

Seating Layout: 3+2 facing.

DTSO(A). Lot No. 31022 (odd nos.) 1987–1988. –/82. 28.2 t.
MSO. Lot No. 31023 1987–1988. –/82. 49.2 t.
TSO. Lot No. 31024 1987–1988. –/77 2T. 31.0 t.
DTSO(B). Lot No. 31025 (even nos.) 1987–1988. –/78. 28.1 t.

319 001	**FU**	P	*FC*	BF	77291	62891	71772	77290
319 002	**FU**	P	*FC*	BF	77293	62892	71773	77292

319 003	**FU**	P	*FC*	BF	77295	62893	71774	77294	
319 004	**FU**	P	*FC*	BF	77297	62894	71775	77296	
319 005	**FU**	P	*FC*	BF	77299	62895	71776	77298	
319 006	**FU**	P	*FC*	BF	77301	62896	71777	77300	
319 007	**FU**	P	*FC*	BF	77303	62897	71778	77302	
319 008	**SN**	P	*FC*	BF	77305	62898	71779	77304	
319 009	**SN**	P	*FC*	BF	77307	62899	71780	77306	
319 010	**FU**	P	*FC*	BF	77309	62900	71781	77308	
319 011	**SN**	P	*FC*	BF	77311	62901	71782	77310	
319 012	**SN**	P	*FC*	BF	77313	62902	71783	77312	
319 013	**SN**	P	*FC*	BF	77315	62903	71784	77314	

Names (carried on TSO):

319 011	John Ruskin College	319 013	The Surrey Hills

Class 319/2. DTSO–MSO–TSO–DTCO. Units converted from Class 319/0 for express services from London to Brighton.

Seating Layout: 1: 2+1 facing, 2: 2+2 facing.

DTSO. Lot No. 31022 (odd nos.) 1987–1988. –/64. 28.2 t.
MSO. Lot No. 31023 1987–1988. –/73 2T. 49.2 t.
TSO. Lot No. 31024 1987–1988. –/52 1T 1TD. 31.0 t.
DTCO. Lot No. 31025 (even nos.) 1987–1988. 18/36. 28.1 t.

319 214	**SN**	P	*FC*	BF	77317	62904	71785	77316	
319 215	**SN**	P	*FC*	BF	77319	62905	71786	77318	London
319 216	**SN**	P	*FC*	BF	77321	62906	71787	77320	
319 217	**SN**	P	*FC*	BF	77323	62907	71788	77322	Brighton
319 218	**SN**	P	*FC*	BF	77325	62908	71789	77324	Croydon
319 219	**SN**	P	*FC*	BF	77327	62909	71790	77326	
319 220	**SN**	P	*FC*	BF	77329	62910	71791	77328	

Class 319/3. DTSO–MSO–TSO–DTSO. Converted from Class 319/1 by replacing First Class seats with Standard Class seats. Used mainly on the Luton–Sutton/Wimbledon routes.

Seating Layout: 3+2 facing.
Dimensions: 19.33 x 2.82 m.

Advertising livery: 319 364 & 319 365 Thameslink Programme (multi-coloured horizontal stripes with pink ends).

DTSO(A). Lot No. 31063 1990. –/70. 29.0 t.
MSO. Lot No. 31064 1990. –/78. 50.6 t.
TSO. Lot No. 31065 1990. –/74 2T. 31.0 t.
DTSO(B). Lot No. 31066 1990. –/75. 29.7 t.

319 361	**FU**	P	*FC*	BF	77459	63043	71929	77458
319 362	**FU**	P	*FC*	BF	77461	63044	71930	77460
319 363	**FU**	P	*FC*	BF	77463	63045	71931	77462
319 364	**AL**	P	*FC*	BF	77465	63046	71932	77464
319 365	**AL**	P	*FC*	BF	77467	63047	71933	77466
319 366	**FU**	P	*FC*	BF	77469	63048	71934	77468
319 367	**FU**	P	*FC*	BF	77471	63049	71935	77470

319 368	**FU**	P	*FC*	BF	77473	63050	71936	77472
319 369	**FU**	P	*FC*	BF	77475	63051	71937	77474
319 370	**FU**	P	*FC*	BF	77477	63052	71938	77476
319 371	**FU**	P	*FC*	BF	77479	63053	71939	77478
319 372	**FU**	P	*FC*	BF	77481	63054	71940	77480
319 373	**FU**	P	*FC*	BF	77483	63055	71941	77482
319 374	**FU**	P	*FC*	BF	77485	63056	71942	77484
319 375	**FU**	P	*FC*	BF	77487	63057	71943	77486
319 376	**FU**	P	*FC*	BF	77489	63058	71944	77488
319 377	**FU**	P	*FC*	BF	77491	63059	71945	77490
319 378	**FU**	P	*FC*	BF	77493	63060	71946	77492
319 379	**FU**	P	*FC*	BF	77495	63061	71947	77494
319 380	**FU**	P	*FC*	BF	77497	63062	71948	77496
319 381	**FU**	P	*FC*	BF	77973	63093	71979	77974
319 382	**FU**	P	*FC*	BF	77975	63094	71980	77976
319 383	**FU**	P	*FC*	BF	77977	63095	71981	77978
319 384	**FU**	P	*FC*	BF	77979	63096	71982	77980
319 385	**FU**	P	*FC*	BF	77981	63097	71983	77982
319 386	**FU**	P	*FC*	BF	77983	63098	71984	77984

Names (carried on TSO):

319 364 Transforming Blackfriars
319 365 Transforming Farringdon
319 374 Bedford Cauldwell TMD

Class 319/4. DTCO–MSO–TSO–DTSO. Converted from Class 319/0. Refurbished with carpets. DTSO(A) converted to composite. Used mainly on the Bedford–Gatwick–Brighton route.

Seating Layout: 1: 2+1 facing 2: 2+2/3+2 facing.

77331–77381. DTCO. Lot No. 31022 (odd nos.) 1987–1988. 12/51. 28.2 t.
77431–77457. DTCO. Lot No. 31038 (odd nos.) 1988. 12/54 12/51. 28.2 t.
62911–62936. MSO. Lot No. 31023 1987–1988. –/74. 49.2 t.
62961–62974. MSO. Lot No. 31039 1988. –/74. 49.2 t.
71792–71817. TSO. Lot No. 31024 1987–1988. –/67 2T. 31.0 t.
71866–71879. TSO. Lot No. 31040 1988. –67 2T. 31.0 t.
77330–77380. DTSO. Lot No. 31025 (even nos.) 1987–1988. –/71 1W. 28.1 t.
77430–77456. DTSO. Lot No. 31041 (even nos.) 1988. –/71 1W. 28.1 t.

319 421	**FU**	P	*FC*	BF	77331	62911	71792	77330
319 422	**FU**	P	*FC*	BF	77333	62912	71793	77332
319 423	**FU**	P	*FC*	BF	77335	62913	71794	77334
319 424	**FU**	P	*FC*	BF	77337	62914	71795	77336
319 425	**FU**	P	*FC*	BF	77339	62915	71796	77338
319 426	**FU**	P	*FC*	BF	77341	62916	71797	77340
319 427	**FU**	P	*FC*	BF	77343	62917	71798	77342
319 428	**FU**	P	*FC*	BF	77345	62918	71799	77344
319 429	**FU**	P	*FC*	BF	77347	62919	71800	77346
319 430	**TL**	P	*FC*	BF	77349	62920	71801	77348
319 431	**FU**	P	*FC*	BF	77351	62921	71802	77350
319 432	**FU**	P	*FC*	BF	77353	62922	71803	77352
319 433	**FU**	P	*FC*	BF	77355	62923	71804	77354

319 434	**FU**	P	*FC*	BF	77357	62924	71805	77356
319 435	**FU**	P	*FC*	BF	77359	62925	71806	77358
319 436	**FU**	P	*FC*	BF	77361	62926	71807	77360
319 437	**FU**	P	*FC*	BF	77363	62927	71808	77362
319 438	**FU**	P	*FC*	BF	77365	62928	71809	77364
319 439	**FU**	P	*FC*	BF	77367	62929	71810	77366
319 440	**FU**	P	*FC*	BF	77369	62930	71811	77368
319 441	**FU**	P	*FC*	BF	77371	62931	71812	77370
319 442	**FU**	P	*FC*	BF	77373	62932	71813	77372
319 443	**FU**	P	*FC*	BF	77375	62933	71814	77374
319 444	**FU**	P	*FC*	BF	77377	62934	71815	77376
319 445	**FU**	P	*FC*	BF	77379	62935	71816	77378
319 446	**FU**	P	*FC*	BF	77381	62936	71817	77380
319 447	**FU**	P	*FC*	BF	77431	62961	71866	77430
319 448	**FU**	P	*FC*	BF	77433	62962	71867	77432
319 449	**FU**	P	*FC*	BF	77435	62963	71868	77434
319 450	**FU**	P	*FC*	BF	77437	62964	71869	77436
319 451	**FU**	P	*FC*	BF	77439	62965	71870	77438
319 452	**FU**	P	*FC*	BF	77441	62966	71871	77440
319 453	**FU**	P	*FC*	BF	77443	62967	71872	77442
319 454	**FU**	P	*FC*	BF	77445	62968	71873	77444
319 455	**FU**	P	*FC*	BF	77447	62969	71874	77446
319 456	**FU**	P	*FC*	BF	77449	62970	71875	77448
319 457	**FU**	P	*FC*	BF	77451	62971	71876	77450
319 458	**FU**	P	*FC*	BF	77453	62972	71877	77452
319 459	**FU**	P	*FC*	BF	77455	62973	71878	77454
319 460	**FU**	P	*FC*	BF	77457	62974	71879	77456

Names (carried on TSO):

319 425 Transforming Travel
319 435 Adrian Jackson-Robbins Chairman 1987–2007 Association of Public Transport Users
319 446 St Pancras International
319 449 King's Cross Thameslink

CLASS 320 BREL YORK

Suburban units.

Formation: DTSO–MSO–DTSO.
Construction: Steel
Traction Motors: Four Brush TM2141B of 268 kW.
Wheel Arrangement: 2-2 + Bo-Bo + 2-2.
Braking: Disc.
Bogies: P7-4 (MSO), T3-7 (others).
Gangways: Within unit.
Doors: Sliding.
Dimensions: 19.33 x 2.82 m.
Couplers: Tightlock.
Control System: Thyristor.
Maximum Speed: 90 m.p.h.
Seating Layout: 3+2 facing.
Multiple Working: Within class & with Classes 317, 318, 319, 321, 322 and 323.

DTSO (A). Lot No. 31060 1990. –/76 1W. 29.1 t.
MSO. Lot No. 31062 1990. –/76 1W. 51.8 t.
DTSO (B). Lot No. 31061 1990. –/75. 30.0 t.

320 301	**SC**	E	*SR*	GW	77899	63021	77921
320 302	**SC**	E	*SR*	GW	77900	63022	77922
320 303	**SC**	E	*SR*	GW	77901	63023	77923
320 304	**SC**	E	*SR*	GW	77902	63024	77924
320 305	**SC**	E	*SR*	GW	77903	63025	77925
320 306	**SC**	E	*SR*	GW	77904	63026	77926
320 307	**SC**	E	*SR*	GW	77905	63027	77927
320 308	**SC**	E	*SR*	GW	77906	63028	77928
320 309	**SC**	E	*SR*	GW	77907	63029	77929
320 310	**SC**	E	*SR*	GW	77908	63030	77930
320 311	**SC**	E	*SR*	GW	77909	63031	77931
320 312	**SC**	E	*SR*	GW	77910	63032	77932
320 313	**SC**	E	*SR*	GW	77911	63033	77933
320 314	**SC**	E	*SR*	GW	77912	63034	77934
320 315	**SC**	E	*SR*	GW	77913	63035	77935
320 316	**SC**	E	*SR*	GW	77914	63036	77936
320 317	**SC**	E	*SR*	GW	77915	63037	77937
320 318	**SC**	E	*SR*	GW	77916	63038	77938
320 319	**SC**	E	*SR*	GW	77917	63039	77939
320 320	**SC**	E	*SR*	GW	77918	63040	77940
320 321	**SC**	E	*SR*	GW	77919	63041	77941
320 322	**SC**	E	*SR*	GW	77920	63042	77942

Names (carried on MSO):

320 305	GLASGOW SCHOOL OF ART 1845 150 1995
320 306	Model Rail Scotland
320 308	High Road 20th Anniversary 2000
320 309	Radio Clyde 25th Anniversary
320 311	Royal College of Physicians and Surgeons of Glasgow
320 312	Sir William A Smith Founder of the Boys' Brigade
320 321	The Rt. Hon. John Smith, QC, MP
320 322	Festive Glasgow Orchid

CLASS 321 BREL YORK

Outer suburban units.

Formation: DTCO (DTSO on Class 321/9)–MSO–TSO–DTSO.
Construction: Steel.
Traction Motors: Four Brush TM2141C (268 kW).
Wheel Arrangement: 2-2 + Bo-Bo + 2-2 + 2-2.
Braking: Disc.
Dimensions: 19.95 x 2.82 m.
Bogies: P7-4 (MSO), T3-7 (others).
Couplers: Tightlock.
Gangways: Within unit.
Control System: Thyristor.
Doors: Sliding.
Maximum Speed: 100 m.p.h.
Seating Layout: 1: 2+2 facing, 2: 3+2 facing.
Multiple Working: Within class & with Classes 317, 318, 319, 320, 322 and 323.

Class 321/3.

DTCO. Lot No. 31053 1988–1990. 16/57. 29.7 t.
MSO. Lot No. 31054 1988–1990. –/82. 51.5 t.
TSO. Lot No. 31055 1988–1990. –/75 2T. 29.1 t.
DTSO. Lot No. 31056 1988–1990. –/78. 29.7 t.

321 301	**NX**	E	*EA*	IL	78049	62975	71880	77853
321 302	**NX**	E	*EA*	IL	78050	62976	71881	77854
321 303	**NX**	E	*EA*	IL	78051	62977	71882	77855
321 304	**NX**	E	*EA*	IL	78052	62978	71883	77856
321 305	**NX**	E	*EA*	IL	78053	62979	71884	77857
321 306	**NX**	E	*EA*	IL	78054	62980	71885	77858
321 307	**NX**	E	*EA*	IL	78055	62981	71886	77859
321 308	**NX**	E	*EA*	IL	78056	62982	71887	77860
321 309	**NX**	E	*EA*	IL	78057	62983	71888	77861
321 310	**NX**	E	*EA*	IL	78058	62984	71889	77862
321 311	**NX**	E	*EA*	IL	78059	62985	71890	77863
321 312	**NX**	E	*EA*	IL	78060	62986	71891	77864
321 313	**NX**	E	*EA*	IL	78061	62987	71892	77865
321 314	**NX**	E	*EA*	IL	78062	62988	71893	77866
321 315	**NX**	E	*EA*	IL	78063	62989	71894	77867
321 316	**NX**	E	*EA*	IL	78064	62990	71895	77868
321 317	**NX**	E	*EA*	IL	78065	62991	71896	77869
321 318	**NX**	E	*EA*	IL	78066	62992	71897	77870
321 319	**NX**	E	*EA*	IL	78067	62993	71898	77871
321 320	**NX**	E	*EA*	IL	78068	62994	71899	77872
321 321	**NX**	E	*EA*	IL	78069	62995	71900	77873
321 322	**NX**	E	*EA*	IL	78070	62996	71901	77874
321 323	**NX**	E	*EA*	IL	78071	62997	71902	77875
321 324	**NX**	E	*EA*	IL	78072	62998	71903	77876
321 325	**NX**	E	*EA*	IL	78073	62999	71904	77877
321 326	**NX**	E	*EA*	IL	78074	63000	71905	77878
321 327	**NC**	E	*EA*	IL	78075	63001	71906	77879
321 328	**NX**	E	*EA*	IL	78076	63002	71907	77880
321 329	**NX**	E	*EA*	IL	78077	63003	71908	77881
321 330	**NC**	E	*EA*	IL	78078	63004	71909	77882
321 331	**NC**	E	*EA*	IL	78079	63005	71910	77883
321 332	**NC**	E	*EA*	IL	78080	63006	71911	77884
321 333	**NC**	E	*EA*	IL	78081	63007	71912	77885
321 334	**NC**	E	*EA*	IL	78082	63008	71913	77886
321 335	**NC**	E	*EA*	IL	78083	63009	71914	77887
321 336	**NC**	E	*EA*	IL	78084	63010	71915	77888
321 337	**GE**	E	*EA*	IL	78085	63011	71916	77889
321 338	**NC**	E	*EA*	IL	78086	63012	71917	77890
321 339	**GE**	E	*EA*	IL	78087	63013	71918	77891
321 340	**GE**	E	*EA*	IL	78088	63014	71919	77892
321 341	**GE**	E	*EA*	IL	78089	63015	71920	77893
321 342	**GE**	E	*EA*	IL	78090	63016	71921	77894
321 343	**GE**	E	*EA*	IL	78091	63017	71922	77895
321 344	**GE**	E	*EA*	IL	78092	63018	71923	77896
321 345	**GE**	E	*EA*	IL	78093	63019	71924	77897

321 346	**GE**	E	*EA*	IL	78094	63020	71925	77898
321 347	**GE**	E	*EA*	IL	78131	63105	71991	78280
321 348	**GE**	E	*EA*	IL	78132	63106	71992	78281
321 349	**GE**	E	*EA*	IL	78133	63107	71993	78282
321 350	**GE**	E	*EA*	IL	78134	63108	71994	78283
321 351	**GE**	E	*EA*	IL	78135	63109	71995	78284
321 352	**GE**	E	*EA*	IL	78136	63110	71996	78285
321 353	**GE**	E	*EA*	IL	78137	63111	71997	78286
321 354	**GE**	E	*EA*	IL	78138	63112	71998	78287
321 355	**GE**	E	*EA*	IL	78139	63113	71999	78288
321 356	**GE**	E	*EA*	IL	78140	63114	72000	78289
321 357	**GE**	E	*EA*	IL	78141	63115	72001	78290
321 358	**GE**	E	*EA*	IL	78142	63116	72002	78291
321 359	**GE**	E	*EA*	IL	78143	63117	72003	78292
321 360	**GE**	E	*EA*	IL	78144	63118	72004	78293
321 361	**GE**	E	*EA*	IL	78145	63119	72005	78294
321 362	**GE**	E	*EA*	IL	78146	63120	72006	78295
321 363	**GE**	E	*EA*	IL	78147	63121	72007	78296
321 364	**GE**	E	*EA*	IL	78148	63122	72008	78297
321 365	**GE**	E	*EA*	IL	78149	63123	72009	78298
321 366	**GE**	E	*EA*	IL	78150	63124	72010	78299

Names (carried on TSO):

321 312	Southend-on-Sea
321 313	University of Essex
321 321	NSPCC ESSEX FULL STOP
321 334	Amsterdam
321 336	GEOFFREY FREEMAN ALLEN
321 343	RSA RAILWAY STUDY ASSOCIATION
321 351	GURKHA
321 361	Phoenix

Class 321/4.

DTCO. Lot No. 31067 1989–1990. 28/40. 29.8 t.
MSO. Lot No. 31068 1989–1990. –/79. 51.6 t.
TSO. Lot No. 31069 1989–1990. –/74 2T. 29.2 t.
DTSO. Lot No. 31070 1989–1990. –/78. 29.8 t.

Note: The DTCOs of 321 438–321 448 units have had 12 First Class seats declassified.

321 401	**FU**	E	*FC*	HE	78095	63063	71949	77943
321 402	**FU**	E	*FC*	HE	78096	63064	71950	77944
321 403	**FU**	E	*FC*	HE	78097	63065	71951	77945
321 404	**FU**	E	*FC*	HE	78098	63066	71952	77946
321 405	**SL**	E	*FC*	HE	78099	63067	71953	77947
321 406	**FU**	E	*FC*	HE	78100	63068	71954	77948
321 407	**SL**	E		HE	78101	63069	71955	77949
321 408	**FU**	E		HE	78102	63070	71956	77950
321 409	**FU**	E		HE	78103	63071	71957	77951
321 410	**SL**	E	*FC*	HE	78104	63072	71958	77952
321 411	**LM**	E	*LM*	NN	78105	63073	71959	77953

321 412	**LM**	E	*LM*	NN	78106	63074	71960	77954
321 413	**SL**	E	*LM*	NN	78107	63075	71961	77955
321 414	**LM**	E	*LM*	NN	78108	63076	71962	77956
321 415	**SL**	E	*LM*	NN	78109	63077	71963	77957
321 416	**LM**	E	*LM*	NN	78110	63078	71964	77958
321 417	**LM**	E	*LM*	NN	78111	63079	71965	77959
321 418	**SL**	E		HE	78112	63080	71968	77962
321 419	**SL**	E		HE	78113	63081	71967	77961
321 420	**SL**	E		HE	78114	63082	71966	77960
321 421	**SL**	E	*EA*	IL	78115	63083	71969	77963
321 422	**SL**	E	*EA*	IL	78116	63084	71970	77964
321 423	**NC**	E	*EA*	IL	78117	63085	71971	77965
321 424	**NX**	E	*EA*	IL	78118	63086	71972	77966
321 425	**SL**	E	*EA*	IL	78119	63087	71973	77967
321 426	**NX**	E	*EA*	IL	78120	63088	71974	77968
321 427	**NX**	E	*EA*	IL	78121	63089	71975	77969
321 428	**NX**	E	*EA*	IL	78122	63090	71976	77970
321 429	**NX**	E	*EA*	IL	78123	63091	71977	77971
321 430	**NX**	E	*EA*	IL	78124	63092	71978	77972
321 431	**NX**	E	*EA*	IL	78151	63125	72011	78300
321 432	**NC**	E	*EA*	IL	78152	63126	72012	78301
321 433	**NC**	E	*EA*	IL	78153	63127	72013	78302
321 434	**NC**	E	*EA*	IL	78154	63128	72014	78303
321 435	**SL**	E	*EA*	IL	78155	63129	72015	78304
321 436	**NC**	E	*EA*	IL	78156	63130	72016	78305
321 437	**SL**	E	*EA*	IL	78157	63131	72017	78306
321 438	**GE**	E	*EA*	IL	78158	63132	72018	78307
321 439	**GE**	E	*EA*	IL	78159	63133	72019	78308
321 440	**GE**	E	*EA*	IL	78160	63134	72020	78309
321 441	**GE**	E	*EA*	IL	78161	63135	72021	78310
321 442	**GE**	E	*EA*	IL	78162	63136	72022	78311
321 443	**GE**	E	*EA*	IL	78125	63099	71985	78274
321 444	**GE**	E	*EA*	IL	78126	63100	71986	78275
321 445	**GE**	E	*EA*	IL	78127	63101	71987	78276
321 446	**1**	E	*EA*	IL	78128	63102	71988	78277
321 447	**GE**	E	*EA*	IL	78129	63103	71989	78278
321 448	**GE**	E	*EA*	IL	78130	63104	71990	78279

Names (carried on TSO):

321 403 Stewart Fleming Signalman King's Cross
321 428 The Essex Commuter
321 444 Essex Lifeboats
321 446 George Mullings

Class 321/9. DTSO(A)–MSO–TSO–DTSO(B).

DTSO(A). Lot No. 31108 1991. –/70(8). 29.2 t.
MSO. Lot No. 31109 1991. –/79. 51.1 t.
TSO. Lot No. 31110 1991. –/74 2T. 29.0 t.
DTSO(B). Lot No. 31111 1991. –/70(7) 1W. 29.2 t.

321 901	**YR**	E	*NO*	NL	77990	63153	72128	77993
321 902	**YR**	E	*NO*	NL	77991	63154	72129	77994
321 903	**YR**	E	*NO*	NL	77992	63155	72130	77995

CLASS 322 BREL YORK

Units built for use on Stansted Airport services, now in use with ScotRail.

Formation: DTSO–MSO–TSO–DTSO.
Construction: Steel.
Traction Motors: Four Brush TM2141C (268 kW).
Wheel Arrangement: 2-2 + Bo-Bo + 2-2 + 2-2.
Braking: Disc.
Bogies: P7-4 (MSO), T3-7 (others).
Gangways: Within unit.
Doors: Sliding.
Dimensions: 19.95/19.92 x 2.82 m.
Couplers: Tightlock.
Control System: Thyristor.
Maximum Speed: 100 m.p.h.
Seating Layout: 3+2 facing.
Multiple Working: Within class & with Classes 317, 318, 319, 320, 321 and 323.

DTSO(A). Lot No. 31094 1990. –/58. 29.3 t.
MSO. Lot No. 31092 1990. –/83. 51.5 t.
TSO. Lot No. 31093 1990. –/76 2T. 28.8 t.
DTSO(B). Lot No. 31091 1990. –/74(2) 1W. 29.1 t.

322 481	**FS**	E	*SR*	GW	78163	63137	72023	77985
322 482	**FS**	E	*SR*	GW	78164	63138	72024	77986
322 483	**FS**	E	*SR*	GW	78165	63139	72025	77987
322 484	**FS**	E	*SR*	GW	78166	63140	72026	77988
322 485	**FS**	E	*SR*	GW	78167	63141	72027	77989

Name (carried on DTSO(A)):

322 481 North Berwick Flyer 1850–2000

CLASS 323 HUNSLET TRANSPORTATION PROJECTS

Suburban units.

Formation: DMSO–PTSO–DMSO.
Construction: Welded aluminium alloy.
Traction Motors: Four Holec DMKT 52/24 asynchronous of 146 kW.
Wheel Arrangement: Bo-Bo + 2-2 + Bo-Bo.
Braking: Disc.
Bogies: SRP BP62 (DMSO), BT52 (PTSO).
Gangways: Within unit.
Doors: Sliding plug.
Dimensions: 23.37/23.44 x 2.80 m.
Couplers: Tightlock.
Control System: GTO Inverter.
Maximum Speed: 90 m.p.h.
Seating Layout: 3+2 facing/unidirectional.

Multiple Working. Within class & with Classes 317, 318, 319, 320, 321 and 322.

DMSO(A). Lot No. 31112 Hunslet 1992–1993. –/98 (* –/82). 41.0 t.
TSO. Lot No. 31113 Hunslet 1992–1993. –/88(5) 1T 2W. (* –/80 1T 2W). 39.4 t.
DMSO(B). Lot No. 31114 Hunslet 1992–1993. –/98 (* –/82). 41.0 t.

323 201		**LM**	P	*LM*	SO	64001	72201	65001
323 202		**LM**	P	*LM*	SO	64002	72202	65002
323 203		**LM**	P	*LM*	SO	64003	72203	65003
323 204		**LM**	P	*LM*	SO	64004	72204	65004
323 205		**LM**	P	*LM*	SO	64005	72205	65005
323 206		**LM**	P	*LM*	SO	64006	72206	65006
323 207		**LM**	P	*LM*	SO	64007	72207	65007
323 208		**LM**	P	*LM*	SO	64008	72208	65008
323 209		**LM**	P	*LM*	SO	64009	72209	65009
323 210		**LM**	P	*LM*	SO	64010	72210	65010
323 211		**LM**	P	*LM*	SO	64011	72211	65011
323 212		**LM**	P	*LM*	SO	64012	72212	65012
323 213		**LM**	P	*LM*	SO	64013	72213	65013
323 214		**LM**	P	*LM*	SO	64014	72214	65014
323 215		**LM**	P	*LM*	SO	64015	72215	65015
323 216		**LM**	P	*LM*	SO	64016	72216	65016
323 217		**LM**	P	*LM*	SO	64017	72217	65017
323 218		**LM**	P	*LM*	SO	64018	72218	65018
323 219		**LM**	P	*LM*	SO	64019	72219	65019
323 220		**LM**	P	*LM*	SO	64020	72220	65020
323 221		**LM**	P	*LM*	SO	64021	72221	65021
323 222		**LM**	P	*LM*	SO	64022	72222	65022
323 223	*	**NO**	P	*NO*	LG	64023	72223	65023
323 224	*	**NO**	P	*NO*	LG	64024	72224	65024
323 225	*	**NO**	P	*NO*	LG	64025	72225	65025
323 226		**NO**	P	*NO*	LG	64026	72226	65026
323 227		**NO**	P	*NO*	LG	64027	72227	65027
323 228		**NO**	P	*NO*	LG	64028	72228	65028
323 229		**NO**	P	*NO*	LG	64029	72229	65029
323 230		**NO**	P	*NO*	LG	64030	72230	65030
323 231		**NO**	P	*NO*	LG	64031	72231	65031
323 232		**NO**	P	*NO*	LG	64032	72232	65032
323 233		**NO**	P	*NO*	LG	64033	72233	65033
323 234		**NO**	P	*NO*	LG	64034	72234	65034
323 235		**NO**	P	*NO*	LG	64035	72235	65035
323 236		**NO**	P	*NO*	LG	64036	72236	65036
323 237		**NO**	P	*NO*	LG	64037	72237	65037
323 238		**NO**	P	*NO*	LG	64038	72238	65038
323 239		**NO**	P	*NO*	LG	64039	72239	65039
323 240		**LM**	P	*LM*	SO	64040	72340	65040
323 241		**LM**	P	*LM*	SO	64041	72341	65041
323 242		**LM**	P	*LM*	SO	64042	72342	65042
323 243		**LM**	P	*LM*	SO	64043	72343	65043

CLASS 325 ABB DERBY

Postal units based on Class 319s. Compatible with diesel or electric locomotive haulage.

Formation: DTPMV–MPMV–TPMV–DTPMV.
System: 25 kV AC overhead/750 V DC third rail.
Construction: Steel.
Traction Motors: Four GEC G315BZ of 268 kW.
Wheel Arrangement: 2-2 + Bo-Bo + 2-2 + 2-2.
Braking: Disc.
Bogies: P7-4 (MSO), T3-7 (others).
Gangways: None.
Doors: Roller shutter.
Multiple Working: Within class.
Dimensions: 19.33 x 2.82 m.
Couplers: Drop-head buckeye.
Control System: GTO Chopper.
Maximum Speed: 100 m.p.h.

DTPMV. Lot No. 31144 1995. 29.1 t.
MPMV. Lot No. 31145 1995. 49.5 t.
TPMV. Lot No. 31146 1995. 30.7 t.

325 001	**RM**	RM *DB*	CE	68300	68340	68360	68301
325 002	**RM**	RM *DB*	CE	68302	68341	68361	68303
325 003	**RM**	RM *DB*	CE	68304	68342	68362	68305
325 004	**RM**	RM *DB*	CE	68306	68343	68363	68307
325 005	**RM**	RM *DB*	CE	68308	68344	68364	68309
325 006	**RM**	RM *DB*	CE	68310	68345	68365	68311
325 007	**RM**	RM *DB*	CE	68312	68346	68366	68313
325 008	**RM**	RM *DB*	CE	68314	68347	68367	68315
325 009	**RM**	RM *DB*	CE	68316	68349	68368	68317
325 010	**RM**	RM	ZI	68318	68348	68369	68319
325 011	**RM**	RM *DB*	CE	68320	68350	68370	68321
325 012	**RM**	RM *DB*	CE	68322	68351	68371	68323
325 013	**RM**	RM *DB*	CE	68324	68352	68372	68325
325 014	**RM**	RM *DB*	CE	68326	68353	68373	68327
325 015	**RM**	RM *DB*	CE	68328	68354	68374	68329
325 016	**RM**	RM *DB*	CE	68330	68355	68375	68331

Names (carried on one side of each DTPMV):

325 002 Royal Mail North Wales & North West
325 006 John Grierson
325 008 Peter Howarth CBE

CLASS 332 HEATHROW EXPRESS SIEMENS

Dedicated Heathrow Express units. Five units were increased from 4-car to 5-car in 2002. Usually operate in coupled pairs.

Formations: Various.
Construction: Steel.
Traction Motors: Two Siemens monomotors asynchronous of 350 kW.
Wheel Arrangement: B-B + 2-2 + 2-2 (+ 2-2) + B-B.

Braking. Disc.
Bogies: CAF.
Gangways: Within unit.
Doors: Sliding plug.
Dimensions: 23.63/23.35 x 2.75 m.
Couplers: Scharfenberg 10L.
Control System: IGBT Inverter.
Maximum Speed: 100 m.p.h.
Heating & ventilation: Air conditioning.
Seating Layout: 1: 2+1 facing, 2: 2+2 mainly unidirectional.
Multiple Working: Within class.

332 001–332 007. DMFO–TSO–PTSO–(TSO)–DMSO.

DMFO. CAF 1997–1998. 26/–. 48.8 t.
72400–72413. TSO. CAF 1997–1998. –/56 35.8 t.
72414–72418. TSO. CAF 2002. –/56 35.8 t.
PTSO. CAF 1997–1998. –/44 1TD 1W. 45.6 t.
DMSO. CAF 1997–1998. –/48. 48.8 t.
DMLFO. CAF 1997–1998. 14/– 1W. 48.8 t.

Advertising livery: Vehicles 78401, 78402, 78405, 78406, 78408, 78410, 78412 Royal Bank of Scotland (deep blue).

332 001	**HE**	HE *HE*	OH	78400	72412	63400		78401
332 002	**HE**	HE *HE*	OH	78402	72409	63401		78403
332 003	**HE**	HE *HE*	OH	78404	72407	63402		78405
332 004	**HE**	HE *HE*	OH	78406	72405	63403		78407
332 005	**HE**	HE *HE*	OH	78408	72411	63404	72417	78409
332 006	**HE**	HE *HE*	OH	78410	72410	63405	72415	78411
332 007	**HE**	HE *HE*	OH	78412	72401	63406	72414	78413

332 008–332 014. DMSO–TSO–PTSO–(TSO)–DMLFO.

Advertising livery: Vehicles 78414, 78416, 78419, 78421, 78423, 78425, 78427 Royal Bank of Scotland (deep blue).

332 008	**HE**	HE *HE*	OH	78414	72413	63407	72418	78415
332 009	**HE**	HE *HE*	OH	78416	72400	63408	72416	78417
332 010	**HE**	HE *HE*	OH	78418	72402	63409		78419
332 011	**HE**	HE *HE*	OH	78420	72403	63410		78421
332 012	**HE**	HE *HE*	OH	78422	72404	63411		78423
332 013	**HE**	HE *HE*	OH	78424	72408	63412		78425
332 014	**HE**	HE *HE*	OH	78426	72406	63413		78427

CLASS 333 SIEMENS

West Yorkshire area suburban units.

Formation: DMSO–PTSO–TSO–DMSO.
Construction: Steel.
Traction Motors: Two Siemens monomotors asynchronous of 350 kW.
Wheel Arrangement: B-B + 2-2 + 2-2 + B-B.
Braking: Disc.
Dimensions: 23.74 (outer ends)/23.35 (TSO) x 2.75 m.
Bogies: CAF.
Gangways: Within unit.
Doors: Sliding plug.
Couplers: Dellner 10L.
Control System: IGBT Inverter.
Maximum Speed: 100 m.p.h.

Heating & ventilation: Air conditioning.
Seating Layout: 3+2 facing/unidirectional.
Multiple Working: Within class.

DMSO(A). (Odd Nos.) CAF 2001. –/90. 50.6 t.
PTSO. CAF 2001. –/73(6) 1TD 2W. 46.0 t.
TSO. CAF 2002–2003. –/100. 38.5 t.
DMSO(B). (Even Nos.) CAF 2001. –/90. 50.0 t.

Notes: 333 001–333 008 were made up to 4-car units from 3-car units in 2002.

333 009–333 016 were made up to 4-car units from 3-car units in 2003.

333 001	**YR**	A	*NO*	NL	78451	74461	74477	78452
333 002	**YR**	A	*NO*	NL	78453	74462	74478	78454
333 003	**YR**	A	*NO*	NL	78455	74463	74479	78456
333 004	**YR**	A	*NO*	NL	78457	74464	74480	78458
333 005	**YR**	A	*NO*	NL	78459	74465	74481	78460
333 006	**YR**	A	*NO*	NL	78461	74466	74482	78462
333 007	**YR**	A	*NO*	NL	78463	74467	74483	78464
333 008	**YR**	A	*NO*	NL	78465	74468	74484	78466
333 009	**YR**	A	*NO*	NL	78467	74469	74485	78468
333 010	**YR**	A	*NO*	NL	78469	74470	74486	78470
333 011	**YR**	A	*NO*	NL	78471	74471	74487	78472
333 012	**YR**	A	*NO*	NL	78473	74472	74488	78474
333 013	**YR**	A	*NO*	NL	78475	74473	74489	78476
333 014	**YR**	A	*NO*	NL	78477	74474	74490	78478
333 015	**YR**	A	*NO*	NL	78479	74475	74491	78480
333 016	**YR**	A	*NO*	NL	78481	74476	74492	78482

Name:

333 007 Alderman J Arthur Godwin First Lord Mayor of Bradford 1907

CLASS 334 JUNIPER ALSTOM BIRMINGHAM

Outer suburban units.

Formation: DMSO–PTSO–DMSO.
Construction: Steel.
Traction Motors: Two Alstom ONIX 800 asynchronous of 270 kW.
Wheel Arrangement: 2-Bo + 2-2 + Bo-2.
Braking: Disc.
Bogies: Alstom LTB3/TBP3.
Gangways: Within unit.
Doors: Sliding plug.
Dimensions: 21.01/19.94 x 2.80 m.
Couplers: Tightlock.
Control System: IGBT Inverter.
Maximum Speed: 90 m.p.h.
Heating & ventilation: Pressure heating and ventilation.
Seating Layout: 2+2 facing/unidirectional (3+2 in PTSO).
Multiple Working: Within class.

64101–64140. DMSO. Alstom Birmingham 1999–2001. –/64. 42.6 t.
PTSO. Alstom Birmingham 1999–2001. –/55 1TD 1W. 39.4 t.
65101–65140. DMSO. Alstom Birmingham 1999–2001. –/64. 42.6 t.

334 001	**SP**	E	*SR*	GW	64101	74301	65101	Donald Dewar
334 002	**SP**	E	*SR*	GW	64102	74302	65102	
334 003	**SP**	E	*SR*	GW	64103	74303	65103	
334 004	**SP**	E	*SR*	GW	64104	74304	65104	
334 005	**SP**	E	*SR*	GW	64105	74305	65105	
334 006	**SP**	E	*SR*	GW	64106	74306	65106	
334 007	**SP**	E	*SR*	GW	64107	74307	65107	
334 008	**SP**	E	*SR*	GW	64108	74308	65108	
334 009	**SP**	E	*SR*	GW	64109	74309	65109	
334 010	**SP**	E	*SR*	GW	64110	74310	65110	
334 011	**SP**	E	*SR*	GW	64111	74311	65111	
334 012	**SP**	E	*SR*	GW	64112	74312	65112	
334 013	**SP**	E	*SR*	GW	64113	74313	65113	
334 014	**SP**	E	*SR*	GW	64114	74314	65114	
334 015	**SP**	E	*SR*	GW	64115	74315	65115	
334 016	**SP**	E	*SR*	GW	64116	74316	65116	
334 017	**SP**	E	*SR*	GW	64117	74317	65117	
334 018	**SP**	E	*SR*	GW	64118	74318	65118	
334 019	**SP**	E	*SR*	GW	64119	74319	65119	
334 020	**SP**	E	*SR*	GW	64120	74320	65120	
334 021	**SP**	E	*SR*	GW	64121	74321	65121	Larkhall
334 022	**SP**	E	*SR*	GW	64122	74322	65122	
334 023	**SP**	E	*SR*	GW	64123	74323	65123	
334 024	**SP**	E	*SR*	GW	64124	74324	65124	
334 025	**SP**	E	*SR*	GW	64125	74325	65125	
334 026	**SP**	E	*SR*	GW	64126	74326	65126	
334 027	**SP**	E	*SR*	GW	64127	74327	65127	
334 028	**SP**	E	*SR*	GW	64128	74328	65128	
334 029	**SP**	E	*SR*	GW	64129	74329	65129	
334 030	**SP**	E	*SR*	GW	64130	74330	65130	
334 031	**SP**	E	*SR*	GW	64131	74331	65131	
334 032	**SP**	E	*SR*	GW	64132	74332	65132	
334 033	**SP**	E	*SR*	GW	64133	74333	65133	
334 034	**SP**	E	*SR*	GW	64134	74334	65134	
334 035	**SP**	E	*SR*	GW	64135	74335	65135	
334 036	**SP**	E	*SR*	GW	64136	74336	65136	
334 037	**SP**	E	*SR*	GW	64137	74337	65137	
334 038	**SP**	E	*SR*	GW	64138	74338	65138	
334 039	**SP**	E	*SR*	GW	64139	74339	65139	
334 040	**SP**	E	*SR*	GW	64140	74340	65140	

CLASS 350 DESIRO UK SIEMENS

Outer suburban and long distance units.

Formation: DMCO–TCO–PTSO–DMCO.
Systems: 25 kV AC overhead (350/1s built with 750 V DC).
Construction: Welded aluminium.
Traction Motors: 4 Siemens 1TB2016-0GB02 asynchronous of 250 kW.
Wheel Arrangement: Bo-Bo + 2-2 + 2-2 + Bo-Bo.

Braking: Disc & regenerative.
Bogies: SGP SF5000.
Gangways: Throughout.
Doors: Sliding plug.
Dimensions: 20.34 x 2.79 m.
Couplers: Dellner 12.
Control System: IGBT Inverter.
Maximum Speed: 100 m.p.h.
Heating & ventilation: Air conditioning.
Seating Layout: 1: 2+2 facing, 2: 2+2 facing/unidirectional (3+2 in 350/2s).
Multiple Working: Within class.

Class 350/1. Original build units owned by Angel Trains. Formerly part of an aborted South West Trains 5-car Class 450/2 order. 2+2 seating.

DMSO(A). Siemens Krefeld 2004–2005. –/60. 48.7 t.
TCO. Siemens Krefeld/Prague 2004–2005. 24/32 1T. 36.2 t.
PTSO. Siemens Krefeld/Prague 2004–2005. –/50(9) 1TD 2W. 45.2 t.
DMSO(B). Siemens Krefeld 2004–2005. –/60. 49.2 t.

350 101	**LM**	A	*LM*	NN	63761	66811	66861	63711
350 102	**LM**	A	*LM*	NN	63762	66812	66862	63712
350 103	**LM**	A	*LM*	NN	63765	66813	66863	63713
350 104	**LM**	A	*LM*	NN	63764	66814	66864	63714
350 105	**LM**	A	*LM*	NN	63763	66815	66868	63715
350 106	**LM**	A	*LM*	NN	63766	66816	66866	63716
350 107	**LM**	A	*LM*	NN	63767	66817	66867	63717
350 108	**LM**	A	*LM*	NN	63768	66818	66865	63718
350 109	**LM**	A	*LM*	NN	63769	66819	66869	63719
350 110	**LM**	A	*LM*	NN	63770	66820	66870	63720
350 111	**LM**	A	*LM*	NN	63771	66821	66871	63721
350 112	**LM**	A	*LM*	NN	63772	66822	66872	63722
350 113	**LM**	A	*LM*	NN	63773	66823	66873	63723
350 114	**LM**	A	*LM*	NN	63774	66824	66874	63724
350 115	**LM**	A	*LM*	NN	63775	66825	66875	63725
350 116	**LM**	A	*LM*	NN	63776	66826	66876	63726
350 117	**LM**	A	*LM*	NN	63777	66827	66877	63727
350 118	**LM**	A	*LM*	NN	63778	66828	66878	63728
350 119	**LM**	A	*LM*	NN	63779	66829	66879	63729
350 120	**LM**	A	*LM*	NN	63780	66830	66880	63730
350 121	**LM**	A	*LM*	NN	63781	66831	66881	63731
350 122	**LM**	A	*LM*	NN	63782	66832	66882	63732
350 123	**LM**	A	*LM*	NN	63783	66833	66883	63733
350 124	**LM**	A	*LM*	NN	63784	66834	66884	63734
350 125	**LM**	A	*LM*	NN	63785	66835	66885	63735
350 126	**LM**	A	*LM*	NN	63786	66836	66886	63736
350 127	**LM**	A	*LM*	NN	63787	66837	66887	63737
350 128	**LM**	A	*LM*	NN	63788	66838	66888	63738
350 129	**LM**	A	*LM*	NN	63789	66839	66889	63739
350 130	**LM**	A	*LM*	NN	63790	66840	66890	63740

Class 350/2. Owned by Porterbrook Leasing. 3+2 seating.

DMSO(A). Siemens Krefeld 2008–2009. –/70. 43.7 t.
TCO. Siemens Prague 2008–2009. 24/42 1T. 35.3 t.
PTSO. Siemens Prague 2008–2009. –/61(9) 1TD 2W. 42.9 t.
DMSO(B). Siemens Krefeld 2008–2009. –/70. 44.2 t.

350 231	**LM**	P	*LM*	NN	61431	65231	67531	61531
350 232	**LM**	P	*LM*	NN	61432	65232	67532	61532
350 233	**LM**	P	*LM*	NN	61433	65233	67533	61533
350 234	**LM**	P	*LM*	NN	61434	65234	67534	61534
350 235	**LM**	P	*LM*	NN	61435	65235	67535	61535
350 236	**LM**	P	*LM*	NN	61436	65236	67536	61536
350 237	**LM**	P	*LM*	NN	61437	65237	67537	61537
350 238	**LM**	P	*LM*	NN	61438	65238	67538	61538
350 239	**LM**	P	*LM*	NN	61439	65239	67539	61539
350 240	**LM**	P	*LM*	NN	61440	65240	67540	61540
350 241	**LM**	P	*LM*	NN	61441	65241	67541	61541
350 242	**LM**	P	*LM*	NN	61442	65242	67542	61542
350 243	**LM**	P	*LM*	NN	61443	65243	67543	61543
350 244	**LM**	P	*LM*	NN	61444	65244	67544	61544
350 245	**LM**	P	*LM*	NN	61445	65245	67545	61545
350 246	**LM**	P	*LM*	NN	61446	65246	67546	61546
350 247	**LM**	P	*LM*	NN	61447	65247	67547	61547
350 248	**LM**	P	*LM*	NN	61448	65248	67548	61548
350 249	**LM**	P	*LM*	NN	61449	65249	67549	61549
350 250	**LM**	P	*LM*	NN	61450	65250	67550	61550
350 251	**LM**	P	*LM*	NN	61451	65251	67551	61551
350 252	**LM**	P	*LM*	NN	61452	65252	67552	61552
350 253	**LM**	P	*LM*	NN	61453	65253	67553	61553
350 254	**LM**	P	*LM*	NN	61454	65254	67554	61554
350 255	**LM**	P	*LM*	NN	61455	65255	67555	61555
350 256	**LM**	P	*LM*	NN	61456	65256	67556	61556
350 257	**LM**	P	*LM*	NN	61457	65257	67557	61557
350 258	**LM**	P	*LM*	NN	61458	65258	67558	61558
350 259	**LM**	P	*LM*	NN	61459	65259	67559	61559
350 260	**LM**	P	*LM*	NN	61460	65260	67560	61560
350 261	**LM**	P	*LM*	NN	61461	65261	67561	61561
350 262	**LM**	P	*LM*	NN	61462	65262	67562	61562
350 263	**LM**	P	*LM*	NN	61463	65263	67563	61563
350 264	**LM**	P	*LM*	NN	61464	65264	67564	61564
350 265	**LM**	P	*LM*	NN	61465	65265	67565	61565
350 266	**LM**	P	*LM*	NN	61466	65266	67566	61566
350 267	**LM**	P	*LM*	NN	61467	65267	67567	61567

CLASS 357 ELECTROSTAR ADTRANZ/BOMBARDIER DERBY

Provision for 750 V DC supply if required.

Formation: DMSO–MSO–PTSO–DMSO.
Construction: Welded aluminium alloy underframe, sides and roof with steel ends. All sections bolted together.
Traction Motors: Two Adtranz asynchronous of 250 kW.
Wheel Arrangement: 2-Bo + 2-Bo + 2-2 + Bo-2.
Braking: Disc & regenerative.
Bogies: Adtranz P3-25/T3-25.
Gangways: Within unit.
Dimensions: 20.40/19.99 x 2.80 m.
Couplers: Tightlock.
Control System: IGBT Inverter.

Doors: Sliding plug. **Maximum Speed**: 100 m.p.h.
Heating & ventilation: Air conditioning.
Seating Layout: 3+2 facing/unidirectional.
Multiple Working: Within class.

Class 357/0. Owned by Porterbrook Leasing.

DMSO(A). Adtranz Derby 1999–2001. –/71. 40.7 t.
MSO. Adtranz Derby 1999–2001. –/78. 36.7 t.
PTSO. Adtranz Derby 1999–2001. –/58(4) 1TD 2W. 39.5 t.
DMSO(B). Adtranz Derby 1999–2001. –/71. 40.7 t.

Advertising livery:
357 010 c2c "green train" (green with purple doors).

357 001	**NC**	P	*C2*	EM	67651	74151	74051	67751
357 002	**NC**	P	*C2*	EM	67652	74152	74052	67752
357 003	**NC**	P	*C2*	EM	67653	74153	74053	67753
357 004	**NC**	P	*C2*	EM	67654	74154	74054	67754
357 005	**NC**	P	*C2*	EM	67655	74155	74055	67755
357 006	**NC**	P	*C2*	EM	67656	74156	74056	67756
357 007	**NC**	P	*C2*	EM	67657	74157	74057	67757
357 008	**NC**	P	*C2*	EM	67658	74158	74058	67758
357 009	**NC**	P	*C2*	EM	67659	74159	74059	67759
357 010	**AL**	P	*C2*	EM	67660	74160	74060	67760
357 011	**NC**	P	*C2*	EM	67661	74161	74061	67761
357 012	**NC**	P	*C2*	EM	67662	74162	74062	67762
357 013	**NC**	P	*C2*	EM	67663	74163	74063	67763
357 014	**NC**	P	*C2*	EM	67664	74164	74064	67764
357 015	**NC**	P	*C2*	EM	67665	74165	74065	67765
357 016	**NC**	P	*C2*	EM	67666	74166	74066	67766
357 017	**NC**	P	*C2*	EM	67667	74167	74067	67767
357 018	**NC**	P	*C2*	EM	67668	74168	74068	67768
357 019	**NC**	P	*C2*	EM	67669	74169	74069	67769
357 020	**C2**	P	*C2*	EM	67670	74170	74070	67770
357 021	**C2**	P	*C2*	EM	67671	74171	74071	67771
357 022	**C2**	P	*C2*	EM	67672	74172	74072	67772
357 023	**C2**	P	*C2*	EM	67673	74173	74073	67773
357 024	**C2**	P	*C2*	EM	67674	74174	74074	67774
357 025	**C2**	P	*C2*	EM	67675	74175	74075	67775
357 026	**C2**	P	*C2*	EM	67676	74176	74076	67776
357 027	**C2**	P	*C2*	EM	67677	74177	74077	67777
357 028	**C2**	P	*C2*	EM	67678	74178	74078	67778
357 029	**C2**	P	*C2*	EM	67679	74179	74079	67779
357 030	**NC**	P	*C2*	EM	67680	74180	74080	67780
357 031	**NC**	P	*C2*	EM	67681	74181	74081	67781
357 032	**NC**	P	*C2*	EM	67682	74182	74082	67782
357 033	**NC**	P	*C2*	EM	67683	74183	74083	67783
357 034	**NC**	P	*C2*	EM	67684	74184	74084	67784
357 035	**NC**	P	*C2*	EM	67685	74185	74085	67785
357 036	**NC**	P	*C2*	EM	67686	74186	74086	67786
357 037	**NC**	P	*C2*	EM	67687	74187	74087	67787
357 038	**NC**	P	*C2*	EM	67688	74188	74088	67788

357 039	**NC**	P	*C2*	EM	67689	74189	74089	67789
357 040	**NC**	P	*C2*	EM	67690	74190	74090	67790
357 041	**NC**	P	*C2*	EM	67691	74191	74091	67791
357 042	**C2**	P	*C2*	EM	67692	74192	74092	67792
357 043	**C2**	P	*C2*	EM	67693	74193	74093	67793
357 044	**NC**	P	*C2*	EM	67694	74194	74094	67794
357 045	**NC**	P	*C2*	EM	67695	74195	74095	67795
357 046	**C2**	P	*C2*	EM	67696	74196	74096	67796

Names (carried on DMSO(A) and DMSO(B) (one plate on each)):

357 001 BARRY FLAXMAN
357 002 ARTHUR LEWIS STRIDE 1841–1922
357 003 JASON LEONARD
357 004 TONY AMOS
357 011 JOHN LOWING
357 028 London, Tilbury & Southend Railway 1854–2004
357 029 THOMAS WHITELEGG 1840–1922
357 030 ROBERT HARBEN WHITELEGG 1871–1957

Class 357/2. Owned by Angel Trains.

DMSO(A). Bombardier Derby 2001–2002. –/71. 40.7 t.
MSO. Bombardier Derby 2001–2002. –/78. 36.7 t.
PTSO. Bombardier Derby 2001–2002. –/58(4) 1TD 2W. 39.5 t.
DMSO(B). Bombardier Derby 2001–2002. –/71. 40.7 t.

357 201	**NC**	A	*C2*	EM	68601	74701	74601	68701
357 202	**NC**	A	*C2*	EM	68602	74702	74602	68702
357 203	**NC**	A	*C2*	EM	68603	74703	74603	68703
357 204	**NC**	A	*C2*	EM	68604	74704	74604	68704
357 205	**NC**	A	*C2*	EM	68605	74705	74605	68705
357 206	**NC**	A	*C2*	EM	68606	74706	74606	68706
357 207	**NC**	A	*C2*	EM	68607	74707	74607	68707
357 208	**NC**	A	*C2*	EM	68608	74708	74608	68708
357 209	**NC**	A	*C2*	EM	68609	74709	74609	68709
357 210	**NC**	A	*C2*	EM	68610	74710	74610	68710
357 211	**NC**	A	*C2*	EM	68611	74711	74611	68711
357 212	**NC**	A	*C2*	EM	68612	74712	74612	68712
357 213	**NC**	A	*C2*	EM	68613	74713	74613	68713
357 214	**NC**	A	*C2*	EM	68614	74714	74614	68714
357 215	**NC**	A	*C2*	EM	68615	74715	74615	68715
357 216	**NC**	A	*C2*	EM	68616	74716	74616	68716
357 217	**NC**	A	*C2*	EM	68617	74717	74617	68717
357 218	**NC**	A	*C2*	EM	68618	74718	74618	68718
357 219	**NC**	A	*C2*	EM	68619	74719	74619	68719
357 220	**NC**	A	*C2*	EM	68620	74720	74620	68720
357 221	**NC**	A	*C2*	EM	68621	74721	74621	68721
357 222	**C2**	A	*C2*	EM	68622	74722	74622	68722
357 223	**C2**	A	*C2*	EM	68623	74723	74623	68723
357 224	**C2**	A	*C2*	EM	68624	74724	74624	68724
357 225	**NC**	A	*C2*	EM	68625	74725	74625	68725
357 226	**C2**	A	*C2*	EM	68626	74726	74626	68726

357 227	**C2**	A	*C2*	EM	68627	74727	74627	68727
357 228	**C2**	A	*C2*	EM	68628	74728	74628	68728

Names (carried on DMSO(A) and DMSO(B) (one plate on each)):

357 201	KEN BIRD
357 202	KENNY MITCHELL
357 203	HENRY PUMFRETT
357 204	DEREK FOWERS
357 205	JOHN D'SILVA
357 206	MARTIN AUNGIER
357 207	JOHN PAGE
357 208	DAVE DAVIS
357 209	JAMES SNELLING
357 213	UPMINSTER I.E.C.C.
357 217	ALLAN BURNELL

CLASS 360/0 DESIRO UK SIEMENS

Outer suburban/express units.

Formation: DMCO–PTSO–TSO–DMCO.
Construction: Welded aluminium.
Traction Motors: 4 Siemens 1TB2016-0GB02 asynchronous of 250 kW.
Wheel Arrangement: Bo-Bo + 2-2 + 2-2 + Bo-Bo.
Braking: Disc & regenerative.
Dimensions: 20.34 x 2.80 m.
Bogies: SGP SF5000.
Couplers: Dellner 12.
Gangways: Within unit.
Control System: IGBT Inverter.
Doors: Sliding plug.
Maximum Speed: 100 m.p.h.
Heating & ventilation: Air conditioning.
Seating Layout: 1: 2+2 facing, 2: 3+2 facing/unidirectional.
Multiple Working: Within class.

DMCO(A). Siemens Krefeld 2002–2003. 8/59. 45.0 t.
PTSO. Siemens Vienna 2002–2003. –/60(9) 1TD 2W. 43.0 t.
TSO. Siemens Vienna 2002–2003. –/78. 35.0 t.
DMCO(B). Siemens Krefeld 2002–2003. 8/59. 45.0 t.

360 101	**FB**	A	*EA*	IL	65551	72551	74551	68551
360 102	**FB**	A	*EA*	IL	65552	72552	74552	68552
360 103	**FB**	A	*EA*	IL	65553	72553	74553	68553
360 104	**FB**	A	*EA*	IL	65554	72554	74554	68554
360 105	**FB**	A	*EA*	IL	65555	72555	74555	68555
360 106	**FB**	A	*EA*	IL	65556	72556	74556	68556
360 107	**FB**	A	*EA*	IL	65557	72557	74557	68557
360 108	**FB**	A	*EA*	IL	65558	72558	74558	68558
360 109	**FB**	A	*EA*	IL	65559	72559	74559	68559
360 110	**FB**	A	*EA*	IL	65560	72560	74560	68560
360 111	**FB**	A	*EA*	IL	65561	72561	74561	68561
360 112	**FB**	A	*EA*	IL	65562	72562	74562	68562
360 113	**FB**	A	*EA*	IL	65563	72563	74563	68563
360 114	**FB**	A	*EA*	IL	65564	72564	74564	68564
360 115	**NX**	A	*EA*	IL	65565	72565	74565	68565
360 116	**FB**	A	*EA*	IL	65566	72566	74566	68566
360 117	**FB**	A	*EA*	IL	65567	72567	74567	68567
360 118	**FB**	A	*EA*	IL	65568	72568	74568	68568
360 119	**FB**	A	*EA*	IL	65569	72569	74569	68569
360 120	**FB**	A	*EA*	IL	65570	72570	74570	68570
360 121	**FB**	A	*EA*	IL	65571	72571	74571	68571

CLASS 360/2 DESIRO UK SIEMENS

4-car Class 350 testbed units rebuilt for use by Heathrow Express on Paddington–Heathrow Airport stopping services ("Heathrow Connect").

Original 4-car sets 360 201–360 204 were made up to 5-cars during 2007 using additional TSOs. A fifth unit (360 205) was delivered in late 2005 as a 5-car set. This set is now dedicated to Terminals 1&3–Terminal 4 shuttle services.

Formation: DMSO–PTSO–TSO–TSO–DMSO.
Construction: Welded aluminium.
Traction Motors: 4 Siemens 1TB2016-0GB02 asynchronous of 250 kW.
Wheel Arrangement: Bo-Bo + 2-2 + 2-2 + 2-2 + Bo-Bo.
Braking: Disc & regenerative.
Dimensions: 20.34 x 2.80 m.
Bogies: SGP SF5000.
Couplers: Dellner 12.
Gangways: Within unit.
Control System: IGBT Inverter.
Doors: Sliding plug.
Maximum Speed: 100 m.p.h.
Heating & ventilation: Air conditioning.
Seating Layout: 3+2 facing/unidirectional.
Multiple Working: Within class.

DMSO(A). Siemens Krefeld 2002–2006. –/63 (* –/54). 44.8 t.
PTSO. Siemens Krefeld 2002–2006. –/57(9) 1TD 2W (* –/48(9) 2W). 44.2 t.
TSO. Siemens Krefeld 2005–2006. –/74 (* –/62). 35.3 t.
TSO. Siemens Krefeld 2002–2006. –/74 (* –/62). 34.1 t.
DMSO(B). Siemens Krefeld 2002–2006. –/63 (* –/54). 44.4 t.

360 201		**HC**	HE *HC*	OH	78431	63421	72431	72421	78441
360 202		**HC**	HE *HC*	OH	78432	63422	72432	72422	78442
360 203		**HC**	HE *HC*	OH	78433	63423	72433	72423	78443
360 204		**HC**	HE *HC*	OH	78434	63424	72434	72424	78444
360 205	*	**HE**	HE *HC*	OH	78435	63425	72435	72425	78445

CLASS 365 NETWORKER EXPRESS ABB YORK

Outer suburban units.

Formations: DMCO–TSO–PTSO–DMCO.
Systems: 25 kV AC overhead but with 750 V DC third rail capability (units marked * were formerly used on DC lines in the South-East).
Construction: Welded aluminium alloy.
Traction Motors: Four GEC-Alsthom G354CX asynchronous of 157 kW.
Wheel Arrangement: Bo-Bo + 2-2 + 2-2 + Bo-Bo.
Braking: Disc & rheostatic.
Dimensions: 20.89/20.06 x 2.81 m.
Bogies: ABB P3-16/T3-16.
Couplers: Tightlock.
Gangways: Within unit.
Control System: GTO Inverter.
Doors: Sliding plug.
Maximum Speed: 100 m.p.h.
Seating Layout: 1: 2+2 facing, 2: 2+2 facing.
Multiple Working: Within class only.

DMCO(A). Lot No. 31133 1994–1995. 12/56. 41.7 t.
TSO. Lot No. 31134 1994–1995. –/65 1TD (* –/64 1TD) 32.9 t.

PTSO. Lot No. 31135 1994–1995. –/68 1T. 34.6 t.
DMCO(B). Lot No. 31136 1994–1995. 12/56. 41.7 t.

Note: Vehicle 65960 of 365 526 is stored at ZC, whilst the others are at ZN.

Advertising liveries:

365 510 Cambridge & Ely; Cathedral cities (blue & white with various images).
365 519 Peterborough; environment capital (blue & white with various images).
365 531 Nelson's County; Norfolk (blue & white with various images).
365 540 Garden cities of Hertfordshire (blue & white with various images).

365 501	*	**FU**	E	*FC*	HE	65894	72241	72240	65935
365 502	*	**FU**	E	*FC*	HE	65895	72243	72242	65936
365 503	*	**FU**	E	*FC*	HE	65896	72245	72244	65937
365 504	*	**FU**	E	*FC*	HE	65897	72247	72246	65938
365 505	*	**FU**	E	*FC*	HE	65898	72249	72248	65939
365 506	*	**FU**	E	*FC*	HE	65899	72251	72250	65940
365 507	*	**FU**	E	*FC*	HE	65900	72253	72252	65941
365 508	*	**FU**	E	*FC*	HE	65901	72255	72254	65942
365 509	*	**FU**	E	*FC*	HE	65902	72257	72256	65943
365 510	*	**AL**	E	*FC*	HE	65903	72259	72258	65944
365 511	*	**FU**	E	*FC*	HE	65904	72261	72260	65945
365 512	*	**FU**	E	*FC*	HE	65905	72263	72262	65946
365 513	*	**FU**	E	*FC*	HE	65906	72265	72264	65947
365 514	*	**FU**	E	*FC*	HE	65907	72267	72266	65948
365 515	*	**FU**	E	*FC*	HE	65908	72269	72268	65949
365 516	*	**FU**	E	*FC*	HE	65909	72271	72270	65950
365 517		**FU**	E	*FC*	HE	65910	72273	72272	65951
365 518		**FU**	E	*FC*	HE	65911	72275	72274	65952
365 519		**AL**	E	*FC*	HE	65912	72277	72276	65953
365 520		**FU**	E	*FC*	HE	65913	72279	72278	65954
365 521		**FU**	E	*FC*	HE	65914	72281	72280	65955
365 522		**FU**	E	*FC*	HE	65915	72283	72282	65956
365 523		**FU**	E	*FC*	HE	65916	72285	72284	65957
365 524		**FU**	E	*FC*	HE	65917	72287	72286	65958
365 525		**FU**	E	*FC*	HE	65918	72289	72288	65959
365 526		**N**	E		ZN	65919	72291	72290	65960
365 527		**FU**	E	*FC*	HE	65920	72293	72292	65961
365 528		**FU**	E	*FC*	HE	65921	72295	72294	65962
365 529		**FU**	E	*FC*	HE	65922	72297	72296	65963
365 530		**FU**	E	*FC*	HE	65923	72299	72298	65964
365 531		**AL**	E	*FC*	HE	65924	72301	72300	65965
365 532		**FU**	E	*FC*	HE	65925	72303	72302	65966
365 533		**FU**	E	*FC*	HE	65926	72305	72304	65967
365 534		**FU**	E	*FC*	HE	65927	72307	72306	65968
365 535		**FU**	E	*FC*	HE	65928	72309	72308	65969
365 536		**FU**	E	*FC*	HE	65929	72311	72310	65970
365 537		**FU**	E	*FC*	HE	65930	72313	72312	65971
365 538		**FU**	E	*FC*	HE	65931	72315	72314	65972
365 539		**FU**	E	*FC*	HE	65932	72317	72316	65973
365 540		**AL**	E	*FC*	HE	65933	72319	72318	65974
365 541		**FU**	E	*FC*	HE	65934	72321	72320	65975

Names (carried on each DMCO).

365 513 Hornsey Depot
365 514 Captain George Vancouver
365 518 The Fenman
365 527 Robert Stripe Passengers' Champion
365 530 The Intalink Partnership promoting integrated transport in Hertfordshire since 1999
365 536 Rufus Barnes Chief Executive of London TravelWatch for 25 years

CLASS 375 ELECTROSTAR ADTRANZ/BOMBARDIER DERBY

Express and outer suburban units.

Formations: Various.
Systems: 25 kV AC overhead/750 V DC third rail (some third rail only with provision for retro-fitting of AC equipment).
Construction: Welded aluminium alloy underframe, sides and roof with steel ends. All sections bolted together.
Traction Motors: Two Adtranz asynchronous of 250 kW.
Wheel Arrangement: 2-Bo (+ 2-Bo) + 2-2 + Bo-2.
Braking: Disc & regenerative.
Dimensions: 20.40/19.99 x 2.80 m.
Bogies: Adtranz P3-25/T3-25.
Couplers: Dellner 12.
Gangways: Throughout.
Control System: IGBT Inverter.
Doors: Sliding plug.
Maximum Speed: 100 m.p.h.
Heating & ventilation: Air conditioning.
Seating Layout: 1: 2+2 facing/unidirectional (seats behind drivers cab in each DMCO). 2: 2+2 facing/unidirectional (except 375/9 – 3+2 facing/unidirectional).
Multiple Working: Within class and with Classes 376, 377 and 378.
Class 375/3. Express units. 750 V DC only. DMCO–TSO–DMCO.

DMCO(A). Bombardier Derby 2001–2002. 12/48. 43.8 t.
TSO. Bombardier Derby 2001–2002. –/56 1TD 2W. 35.5 t.
DMCO(B). Bombardier Derby 2001–2002. 12/48. 43.8 t.

375 301	**CN**	E	*SE*	RM	67921	74351	67931
375 302	**CN**	E	*SE*	RM	67922	74352	67932
375 303	**CN**	E	*SE*	RM	67923	74353	67933
375 304	**CN**	E	*SE*	RM	67924	74354	67934
375 305	**CN**	E	*SE*	RM	67925	74355	67935
375 306	**CN**	E	*SE*	RM	67926	74356	67936
375 307	**CN**	E	*SE*	RM	67927	74357	67937
375 308	**CN**	E	*SE*	RM	67928	74358	67938
375 309	**CN**	E	*SE*	RM	67929	74359	67939
375 310	**CN**	E	*SE*	RM	67930	74360	67940

Name (carried on TSO):

375 304 Medway Valley Line 1856–2006

Class 375/6. Express units. 25 kV AC/750 V DC. DMCO–MSO–PTSO–DMCO.

DMCO(A). Adtranz Derby 1999–2001. 12/48. 46.2 t.

MSO. Adtranz Derby 1999–2001. –/66 1T. 40.5 t.
PTSO. Adtranz Derby 1999–2001. –/56 1TD 2W. 40.7 t.
DMCO(B). Adtranz Derby 1999–2001. 12/48. 46.2 t.

375 601	**CN**	E	*SE*	RM	67801	74251	74201	67851
375 602	**CN**	E	*SE*	RM	67802	74252	74202	67852
375 603	**CN**	E	*SE*	RM	67803	74253	74203	67853
375 604	**CN**	E	*SE*	RM	67804	74254	74204	67854
375 605	**CN**	E	*SE*	RM	67805	74255	74205	67855
375 606	**CN**	E	*SE*	RM	67806	74256	74206	67856
375 607	**CN**	E	*SE*	RM	67807	74257	74207	67857
375 608	**CN**	E	*SE*	RM	67808	74258	74208	67858
375 609	**CN**	E	*SE*	RM	67809	74259	74209	67859
375 610	**CN**	E	*SE*	RM	67810	74260	74210	67860
375 611	**CN**	E	*SE*	RM	67811	74261	74211	67861
375 612	**CN**	E	*SE*	RM	67812	74262	74212	67862
375 613	**CN**	E	*SE*	RM	67813	74263	74213	67863
375 614	**CN**	E	*SE*	RM	67814	74264	74214	67864
375 615	**CN**	E	*SE*	RM	67815	74265	74215	67865
375 616	**CN**	E	*SE*	RM	67816	74266	74216	67866
375 617	**CN**	E	*SE*	RM	67817	74267	74217	67867
375 618	**CN**	E	*SE*	RM	67818	74268	74218	67868
375 619	**CN**	E	*SE*	RM	67819	74269	74219	67869
375 620	**CN**	E	*SE*	RM	67820	74270	74220	67870
375 621	**CN**	E	*SE*	RM	67821	74271	74221	67871
375 622	**CN**	E	*SE*	RM	67822	74272	74222	67872
375 623	**CN**	E	*SE*	RM	67823	74273	74223	67873
375 624	**CN**	E	*SE*	RM	67824	74274	74224	67874
375 625	**CN**	E	*SE*	RM	67825	74275	74225	67875
375 626	**CN**	E	*SE*	RM	67826	74276	74226	67876
375 627	**CN**	E	*SE*	RM	67827	74277	74227	67877
375 628	**CN**	E	*SE*	RM	67828	74278	74228	67878
375 629	**CN**	E	*SE*	RM	67829	74279	74229	67879
375 630	**CN**	E	*SE*	RM	67830	74280	74230	67880

Names (carried on one side of each MSO or PTSO):

375 608 Bromley Travelwise
375 610 Royal Tunbridge Wells
375 611 Dr. William Harvey
375 619 Driver John Neve
375 623 Hospice in the Weald

Class 375/7. Express units. 750 V DC only. DMCO–MSO–TSO–DMCO.

DMCO(A). Bombardier Derby 2001–2002. 12/48. 43.8 t.
MSO. Bombardier Derby 2001–2002. –/66 1T. 36.4 t.
TSO. Bombardier Derby 2001–2002. –/56 1TD 2W. 34.1 t.
DMCO(B). Bombardier Derby 2001–2002. 12/48. 43.8 t.

375 701	**CN**	E	*SE*	RM	67831	74281	74231	67881
375 702	**CN**	E	*SE*	RM	67832	74282	74232	67882
375 703	**CN**	E	*SE*	RM	67833	74283	74233	67883
375 704	**CN**	E	*SE*	RM	67834	74284	74234	67884
375 705	**CN**	E	*SE*	RM	67835	74285	74235	67885
375 706	**CN**	E	*SE*	RM	67836	74286	74236	67886

375 707	**CN**	E	*SE*	RM	67837	74287	74237	67887
375 708	**CN**	E	*SE*	RM	67838	74288	74238	67888
375 709	**CN**	E	*SE*	RM	67839	74289	74239	67889
375 710	**CN**	E	*SE*	RM	67840	74290	74240	67890
375 711	**CN**	E	*SE*	RM	67841	74291	74241	67891
375 712	**CN**	E	*SE*	RM	67842	74292	74242	67892
375 713	**CN**	E	*SE*	RM	67843	74293	74243	67893
375 714	**CN**	E	*SE*	RM	67844	74294	74244	67894
375 715	**CN**	E	*SE*	RM	67845	74295	74245	67895

Names (carried on one side of each MSO or TSO):

375 701 Kent Air Ambulance Explorer

Class 375/8. Express units. 750 V DC only. DMCO–MSO–TSO–DMCO.

DMCO(A). Bombardier Derby 2004. 12/48. 43.3 t.
MSO. Bombardier Derby 2004. –/66 1T. 39.8 t.
TSO. Bombardier Derby 2004. –/52 1TD 2W. 35.9 t.
DMCO(B). Bombardier Derby 2004. 12/52. 43.3 t.

375 801	**CN**	E	*SE*	RM	73301	79001	78201	73701
375 802	**CN**	E	*SE*	RM	73302	79002	78202	73702
375 803	**CN**	E	*SE*	RM	73303	79003	78203	73703
375 804	**CN**	E	*SE*	RM	73304	79004	78204	73704
375 805	**CN**	E	*SE*	RM	73305	79005	78205	73705
375 806	**CN**	E	*SE*	RM	73306	79006	78206	73706
375 807	**CN**	E	*SE*	RM	73307	79007	78207	73707
375 808	**CN**	E	*SE*	RM	73308	79008	78208	73708
375 809	**CN**	E	*SE*	RM	73309	79009	78209	73709
375 810	**CN**	E	*SE*	RM	73310	79010	78210	73710
375 811	**CN**	E	*SE*	RM	73311	79011	78211	73711
375 812	**CN**	E	*SE*	RM	73312	79012	78212	73712
375 813	**CN**	E	*SE*	RM	73313	79013	78213	73713
375 814	**CN**	E	*SE*	RM	73314	79014	78214	73714
375 815	**CN**	E	*SE*	RM	73315	79015	78215	73715
375 816	**CN**	E	*SE*	RM	73316	79016	78216	73716
375 817	**CN**	E	*SE*	RM	73317	79017	78217	73717
375 818	**CN**	E	*SE*	RM	73318	79018	78218	73718
375 819	**CN**	E	*SE*	RM	73319	79019	78219	73719
375 820	**CN**	E	*SE*	RM	73320	79020	78220	73720
375 821	**CN**	E	*SE*	RM	73321	79021	78221	73721
375 822	**CN**	E	*SE*	RM	73322	79022	78222	73722
375 823	**CN**	E	*SE*	RM	73323	79023	78223	73723
375 824	**CN**	E	*SE*	RM	73324	79024	78224	73724
375 825	**CN**	E	*SE*	RM	73325	79025	78225	73725
375 826	**CN**	E	*SE*	RM	73326	79026	78226	73726
375 827	**CN**	E	*SE*	RM	73327	79027	78227	73727
375 828	**CN**	E	*SE*	RM	73328	79028	78228	73728
375 829	**CN**	E	*SE*	RM	73329	79029	78229	73729
375 830	**CN**	E	*SE*	RM	73330	79030	78230	73730

Name (carried on one side of each MSO or TSO):

375 830 City of London

Class 375/9. Outer suburban units. 750 V DC only. DMCO–MSO–TSO–DMCO.

DMCO(A). Bombardier Derby 2003–2004. 12/59. 43.4 t.
MSO. Bombardier Derby 2003–2004. –/73 1T. 39.3 t.
TSO. Bombardier Derby 2003–2004. –/59 1TD 2W. 35.6 t.
DMCO(B). Bombardier Derby 2003–2004. 12/59. 43.4 t.

375 901	**CN**	E	*SE*	RM	73331	79031	79061	73731
375 902	**CN**	E	*SE*	RM	73332	79032	79062	73732
375 903	**CN**	E	*SE*	RM	73333	79033	79063	73733
375 904	**CN**	E	*SE*	RM	73334	79034	79064	73734
375 905	**CN**	E	*SE*	RM	73335	79035	79065	73735
375 906	**CN**	E	*SE*	RM	73336	79036	79066	73736
375 907	**CN**	E	*SE*	RM	73337	79037	79067	73737
375 908	**CN**	E	*SE*	RM	73338	79038	79068	73738
375 909	**CN**	E	*SE*	RM	73339	79039	79069	73739
375 910	**CN**	E	*SE*	RM	73340	79040	79070	73740
375 911	**CN**	E	*SE*	RM	73341	79041	79071	73741
375 912	**CN**	E	*SE*	RM	73342	79042	79072	73742
375 913	**CN**	E	*SE*	RM	73343	79043	79073	73743
375 914	**CN**	E	*SE*	RM	73344	79044	79074	73744
375 915	**CN**	E	*SE*	RM	73345	79045	79075	73745
375 916	**CN**	E	*SE*	RM	73346	79046	79076	73746
375 917	**CN**	E	*SE*	RM	73347	79047	79077	73747
375 918	**CN**	E	*SE*	RM	73348	79048	79078	73748
375 919	**CN**	E	*SE*	RM	73349	79049	79079	73749
375 920	**CN**	E	*SE*	RM	73350	79050	79080	73750
375 921	**CN**	E	*SE*	RM	73351	79051	79081	73751
375 922	**CN**	E	*SE*	RM	73352	79052	79082	73752
375 923	**CN**	E	*SE*	RM	73353	79053	79083	73753
375 924	**CN**	E	*SE*	RM	73354	79054	79084	73754
375 925	**CN**	E	*SE*	RM	73355	79055	79085	73755
375 926	**CN**	E	*SE*	RM	73356	79056	79086	73756
375 927	**CN**	E	*SE*	RM	73357	79057	79087	73757

CLASS 376 ELECTROSTAR BOMBARDIER DERBY

Inner suburban units.

Formation: DMSO–MSO–TSO–MSO–DMSO.
System: 750 V DC third rail.
Construction: Welded aluminium alloy underframe, sides and roof with steel ends. All sections bolted together.
Traction Motors: Two Bombardier asynchronous of 250 kW.
Wheel Arrangement: 2-Bo + 2-Bo + 2-2 + Bo-2 + Bo-2.
Braking: Disc & regenerative.
Bogies: Bombardier P3-25/T3-25.
Gangways: Within unit.
Doors: Sliding.
Dimensions: 20.40/19.99 x 2.80 m.
Couplers: Dellner 12.
Control System: IGBT Inverter.
Maximum Speed: 75 m.p.h.
Heating & ventilation: Pressure heating and ventilation.
Seating Layout: 2+2 low density facing.
Multiple Working: Within class and with Classes 375, 377 and 378.

DMSO(A). Bombardier Derby 2004–2005. –/36(6) 1W. 42.1 t.
MSO. Bombardier Derby 2004–2005. –/48. 36.2 t.
TSO. Bombardier Derby 2004–2005. –/48. 36.3 t.
DMSO(B). Bombardier Derby 2004–2005. –/36(6) 1W. 42.1 t.

376 001	**CN**	E	*SE*	SG	61101	63301	64301	63501	61601
376 002	**CN**	E	*SE*	SG	61102	63302	64302	63502	61602
376 003	**CN**	E	*SE*	SG	61103	63303	64303	63503	61603
376 004	**CN**	E	*SE*	SG	61104	63304	64304	63504	61604
376 005	**CN**	E	*SE*	SG	61105	63305	64305	63505	61605
376 006	**CN**	E	*SE*	SG	61106	63306	64306	63506	61606
376 007	**CN**	E	*SE*	SG	61107	63307	64307	63507	61607
376 008	**CN**	E	*SE*	SG	61108	63308	64308	63508	61608
376 009	**CN**	E	*SE*	SG	61109	63309	64309	63509	61609
376 010	**CN**	E	*SE*	SG	61110	63310	64310	63510	61610
376 011	**CN**	E	*SE*	SG	61111	63311	64311	63511	61611
376 012	**CN**	E	*SE*	SG	61112	63312	64312	63512	61612
376 013	**CN**	E	*SE*	SG	61113	63313	64313	63513	61613
376 014	**CN**	E	*SE*	SG	61114	63314	64314	63514	61614
376 015	**CN**	E	*SE*	SG	61115	63315	64315	63515	61615
376 016	**CN**	E	*SE*	SG	61116	63316	64316	63516	61616
376 017	**CN**	E	*SE*	SG	61117	63317	64317	63517	61617
376 018	**CN**	E	*SE*	SG	61118	63318	64318	63518	61618
376 019	**CN**	E	*SE*	SG	61119	63319	64319	63519	61619
376 020	**CN**	E	*SE*	SG	61120	63320	64320	63520	61620
376 021	**CN**	E	*SE*	SG	61121	63321	64321	63521	61621
376 022	**CN**	E	*SE*	SG	61122	63322	64322	63522	61622
376 023	**CN**	E	*SE*	SG	61123	63323	64323	63523	61623
376 024	**CN**	E	*SE*	SG	61124	63324	64324	63524	61624
376 025	**CN**	E	*SE*	SG	61125	63325	64325	63525	61625
376 026	**CN**	E	*SE*	SG	61126	63326	64326	63526	61626
376 027	**CN**	E	*SE*	SG	61127	63327	64327	63527	61627
376 028	**CN**	E	*SE*	SG	61128	63328	64328	63528	61628
376 029	**CN**	E	*SE*	SG	61129	63329	64329	63529	61629
376 030	**CN**	E	*SE*	SG	61130	63330	64330	63530	61630
376 031	**CN**	E	*SE*	SG	61131	63331	64331	63531	61631
376 032	**CN**	E	*SE*	SG	61132	63332	64332	63532	61632
376 033	**CN**	E	*SE*	SG	61133	63333	64333	63533	61633
376 034	**CN**	E	*SE*	SG	61134	63334	64334	63534	61634
376 035	**CN**	E	*SE*	SG	61135	63335	64335	63535	61635
376 036	**CN**	E	*SE*	SG	61136	63336	64336	63536	61636

CLASS 377 ELECTROSTAR BOMBARDIER DERBY

Express and outer suburban units.

Formations: Various.
Systems: 25 kV AC overhead/750 V DC third rail or third rail only with provision for retro-fitting of AC equipment.
Construction: Welded aluminium alloy underframe, sides and roof with steel ends. All sections bolted together.
Traction Motors: Two Bombardier asynchronous of 250 kW.

Wheel Arrangement: 2-Bo (+ 2-Bo) + 2-2 + Bo-2.
Braking: Disc & regenerative.
Dimensions: 20.39/19.99 x 2.80 m.
Bogies: Bombardier P3-25/T3-25.
Couplers: Dellner 12.
Gangways: Throughout.
Control System: IGBT Inverter.
Doors: Sliding plug.
Maximum Speed: 100 m.p.h.
Heating & ventilation: Air conditioning.
Seating Layout: Various.
Multiple Working: Within class and with Classes 375, 376 and 378.

Class 377/1. 750 V DC only. DMCO–MSO–TSO–DMCO.
Seating layout: 1: 2+2 facing/unidirectional, 2: 2+2 facing/unidirectional (377 101–377 119), 3+2 and 2+2 facing/unidirectional (377 120–377 164) (3+2 seating in middle cars only 377 140–377 164).

DMCO(A). Bombardier Derby 2002–2003. 12/48 (s 12/56). 44.8 t.
MSO. Bombardier Derby 2002–2003. –/62 (s –/70, t –/69). 1T. 39.0 t.
TSO. Bombardier Derby 2002–2003. –/52 (s –/60, t –/57). 1TD 2W. 35.4 t.
DMCO(B). Bombardier Derby 2002–2003. 12/48 (s 12/56). 43.4 t.

377 101		**SN**	P	*SN*	BI	78501	77101	78901	78701
377 102		**SN**	P	*SN*	BI	78502	77102	78902	78702
377 103		**SN**	P	*SN*	BI	78503	77103	78903	78703
377 104		**SN**	P	*SN*	BI	78504	77104	78904	78704
377 105		**SN**	P	*SN*	BI	78505	77105	78905	78705
377 106		**SN**	P	*SN*	BI	78506	77106	78906	78706
377 107		**SN**	P	*SN*	BI	78507	77107	78907	78707
377 108		**SN**	P	*SN*	BI	78508	77108	78908	78708
377 109		**SN**	P	*SN*	BI	78509	77109	78909	78709
377 110		**SN**	P	*SN*	BI	78510	77110	78910	78710
377 111		**SN**	P	*SN*	BI	78511	77111	78911	78711
377 112		**SN**	P	*SN*	BI	78512	77112	78912	78712
377 113		**SN**	P	*SN*	BI	78513	77113	78913	78713
377 114		**SN**	P	*SN*	BI	78514	77114	78914	78714
377 115		**SN**	P	*SN*	BI	78515	77115	78915	78715
377 116		**SN**	P	*SN*	BI	78516	77116	78916	78716
377 117		**SN**	P	*SN*	BI	78517	77117	78917	78717
377 118		**SN**	P	*SN*	BI	78518	77118	78918	78718
377 119		**SN**	P	*SN*	BI	78519	77119	78919	78719
377 120	s	**SN**	P	*SN*	SU	78520	77120	78920	78720
377 121	s	**SN**	P	*SN*	SU	78521	77121	78921	78721
377 122	s	**SN**	P	*SN*	SU	78522	77122	78922	78722
377 123	s	**SN**	P	*SN*	SU	78523	77123	78923	78723
377 124	s	**SN**	P	*SN*	SU	78524	77124	78924	78724
377 125	s	**SN**	P	*SN*	SU	78525	77125	78925	78725
377 126	s	**SN**	P	*SN*	SU	78526	77126	78926	78726
377 127	s	**SN**	P	*SN*	SU	78527	77127	78927	78727
377 128	s	**SN**	P	*SN*	SU	78528	77128	78928	78728
377 129	s	**SN**	P	*SN*	SU	78529	77129	78929	78729
377 130	s	**SN**	P	*SN*	SU	78530	77130	78930	78730
377 131	s	**SN**	P	*SN*	SU	78531	77131	78931	78731
377 132	s	**SN**	P	*SN*	SU	78532	77132	78932	78732
377 133	s	**SN**	P	*SN*	SU	78533	77133	78933	78733

377 134	s	**SN**	P	*SN*	SU	78534	77134	78034	78734
377 135	s	**SN**	P	*SN*	SU	78535	77135	78935	78735
377 136	s	**SN**	P	*SN*	SU	78536	77136	78936	78736
377 137	s	**SN**	P	*SN*	SU	78537	77137	78937	78737
377 138	s	**SN**	P	*SN*	SU	78538	77138	78938	78738
377 139	s	**SN**	P	*SN*	SU	78539	77139	78939	78739
377 140	t	**SN**	P	*SN*	SU	78540	77140	78940	78740
377 141	t	**SN**	P	*SN*	SU	78541	77141	78941	78741
377 142	t	**SN**	P	*SN*	SU	78542	77142	78942	78742
377 143	t	**SN**	P	*SN*	SU	78543	77143	78943	78743
377 144	t	**SN**	P	*SN*	SU	78544	77144	78944	78744
377 145	t	**SN**	P	*SN*	SU	78545	77145	78945	78745
377 146	t	**SN**	P	*SN*	SU	78546	77146	78946	78746
377 147	t	**SN**	P	*SN*	SU	78547	77147	78947	78747
377 148	t	**SN**	P	*SN*	SU	78548	77148	78948	78748
377 149	t	**SN**	P	*SN*	SU	78549	77149	78949	78749
377 150	t	**SN**	P	*SN*	SU	78550	77150	78950	78750
377 151	t	**SN**	P	*SN*	SU	78551	77151	78951	78751
377 152	t	**SN**	P	*SN*	SU	78552	77152	78952	78752
377 153	t	**SN**	P	*SN*	SU	78553	77153	78953	78753
377 154	t	**SN**	P	*SN*	SU	78554	77154	78954	78754
377 155	t	**SN**	P	*SN*	SU	78555	77155	78955	78755
377 156	t	**SN**	P	*SN*	SU	78556	77156	78956	78756
377 157	t	**SN**	P	*SN*	SU	78557	77157	78957	78757
377 158	t	**SN**	P	*SN*	SU	78558	77158	78958	78758
377 159	t	**SN**	P	*SN*	SU	78559	77159	78959	78759
377 160	t	**SN**	P	*SN*	SU	78560	77160	78960	78760
377 161	t	**SN**	P	*SN*	SU	78561	77161	78961	78761
377 162	t	**SN**	P	*SN*	BI	78562	77162	78962	78762
377 163	t	**SN**	P	*SN*	BI	78563	77163	78963	78763
377 164	t	**SN**	P	*SN*	BI	78564	77164	78964	78764

Class 377/2. 25 kV AC/750 V DC. DMCO–MSO–PTSO–DMCO. These dual-voltage units are used on the East Croydon–Milton Keynes cross-London service.
Seating layout: 1: 2+2 facing/unidirectional, 2: 2+2 and 3+2 facing/unidirectional (3+2 seating in middle cars only).

DMCO(A). Bombardier Derby 2003–2004. 12/48. 44.2 t.
MSO. Bombardier Derby 2003–2004. –/69 1T. 39.8 t.
PTSO. Bombardier Derby 2003–2004. –/57 1TD 2W. 40.1 t.
DMCO(B). Bombardier Derby 2003–2004. 12/48. 44.2 t.

377 201	**SN**	P	*SN*	BI	78571	77171	78971	78771
377 202	**SN**	P	*SN*	SU	78572	77172	78972	78772
377 203	**SN**	P	*SN*	BI	78573	77173	78973	78773
377 204	**SN**	P	*SN*	BI	78574	77174	78974	78774
377 205	**SN**	P	*SN*	SU	78575	77175	78975	78775
377 206	**SN**	P	*SN*	BI	78576	77176	78976	78776
377 207	**SN**	P	*SN*	BI	78577	77177	78977	78777
377 208	**SN**	P	*SN*	SU	78578	77178	78978	78778
377 209	**SN**	P	*SN*	SU	78579	77179	78979	78779
377 210	**SN**	P	*SN*	SU	78580	77180	78980	78780

377 211	**SN**	P	*SN*	SU	78581	77181	78981	78781
377 212	**SN**	P	*SN*	SU	78582	77182	78982	78782
377 213	**SN**	P	*SN*	SU	78583	77183	78983	78783
377 214	**SN**	P	*SN*	SU	78584	77184	78984	78784
377 215	**SN**	P	*SN*	SU	78585	77185	78985	78785

Class 377/3. 750 V DC only. DMCO–TSO–DMCO.
Seating Layout: 1: 2+2 facing/unidirectional, 2: 2+2 facing/unidirectional.

Notes: Units built as Class 375, but renumbered in the Class 377/3 range when fitted with Dellner couplers.

† Wi-fi high-speed internet connection equipment fitted. Units generally used on Victoria–Brighton fast services.

DMCO(A). Bombardier Derby 2001–2002. 12/48. 43.5 t.
TSO. Bombardier Derby 2001–2002. –/56 1TD 2W. 35.4 t.
DMCO(B). Bombardier Derby 2001–2002. 12/48. 43.5 t.

377 301	(375 311)		**SN**	P	*SN*	BI	68201	74801	68401
377 302	(375 312)		**SN**	P	*SN*	BI	68202	74802	68402
377 303	(375 313)		**SN**	P	*SN*	BI	68203	74803	68403
377 304	(375 314)	†	**SN**	P	*SN*	BI	68204	74804	68404
377 305	(375 315)	†	**SN**	P	*SN*	BI	68205	74805	68405
377 306	(375 316)		**SN**	P	*SN*	BI	68206	74806	68406
377 307	(375 317)		**SN**	P	*SN*	BI	68207	74807	68407
377 308	(375 318)		**SN**	P	*SN*	BI	68208	74808	68408
377 309	(375 319)		**SN**	P	*SN*	BI	68209	74809	68409
377 310	(375 320)		**SN**	P	*SN*	BI	68210	74810	68410
377 311	(375 321)		**SN**	P	*SN*	BI	68211	74811	68411
377 312	(375 322)		**SN**	P	*SN*	BI	68212	74812	68412
377 313	(375 323)		**SN**	P	*SN*	BI	68213	74813	68413
377 314	(375 324)		**SN**	P	*SN*	BI	68214	74814	68414
377 315	(375 325)	†	**SN**	P	*SN*	BI	68215	74815	68415
377 316	(375 326)	†	**SN**	P	*SN*	BI	68216	74816	68416
377 317	(375 327)	†	**SN**	P	*SN*	BI	68217	74817	68417
377 318	(375 328)		**SN**	P	*SN*	BI	68218	74818	68418
377 319	(375 329)		**SN**	P	*SN*	BI	68219	74819	68419
377 320	(375 330)	†	**SN**	P	*SN*	BI	68220	74820	68420
377 321	(375 331)	†	**SN**	P	*SN*	BI	68221	74821	68421
377 322	(375 332)	†	**SN**	P	*SN*	BI	68222	74822	68422
377 323	(375 333)		**SN**	P	*SN*	BI	68223	74823	68423
377 324	(375 334)	†	**SN**	P	*SN*	BI	68224	74824	68424
377 325	(375 335)	†	**SN**	P	*SN*	BI	68225	74825	68425
377 326	(375 336)	†	**SN**	P	*SN*	BI	68226	74826	68426
377 327	(375 337)	†	**SN**	P	*SN*	BI	68227	74827	68427
377 328	(375 338)	†	**SN**	P	*SN*	BI	68228	74828	68428

Class 377/4. 750 V DC only. DMCO–MSO–TSO–DMCO.
Seating Layout: 1: 2+2 facing/two seats longitudinal, 2: 2+2 and 3+2 facing/unidirectional (3+2 seating in middle cars only).

DMCO(A). Bombardier Derby 2004–2005. 10/48. 43.1 t.
MSO. Bombardier Derby 2004–2005. –/69 1T. 39.3 t.

TSO. Bombardier Derby 2004–2005. /56 1TD 2W. 35.3 t.
DMCO(B). Bombardier Derby 2004–2005. 10/48. 43.2 t.

377 401	**SN**	P	*SN*	BI	73401	78801	78601	73801
377 402	**SN**	P	*SN*	BI	73402	78802	78602	73802
377 403	**SN**	P	*SN*	BI	73403	78803	78603	73803
377 404	**SN**	P	*SN*	BI	73404	78804	78604	73804
377 405	**SN**	P	*SN*	BI	73405	78805	78605	73805
377 406	**SN**	P	*SN*	BI	73406	78806	78606	73806
377 407	**SN**	P	*SN*	BI	73407	78807	78607	73807
377 408	**SN**	P	*SN*	BI	73408	78808	78608	73808
377 409	**SN**	P	*SN*	BI	73409	78809	78609	73809
377 410	**SN**	P	*SN*	BI	73410	78810	78610	73810
377 411	**SN**	P	*SN*	BI	73411	78811	78611	73811
377 412	**SN**	P	*SN*	BI	73412	78812	78612	73812
377 413	**SN**	P	*SN*	BI	73413	78813	78613	73813
377 414	**SN**	P	*SN*	BI	73414	78814	78614	73814
377 415	**SN**	P	*SN*	BI	73415	78815	78615	73815
377 416	**SN**	P	*SN*	BI	73416	78816	78616	73816
377 417	**SN**	P	*SN*	BI	73417	78817	78617	73817
377 418	**SN**	P	*SN*	BI	73418	78818	78618	73818
377 419	**SN**	P	*SN*	BI	73419	78819	78619	73819
377 420	**SN**	P	*SN*	BI	73420	78820	78620	73820
377 421	**SN**	P	*SN*	BI	73421	78821	78621	73821
377 422	**SN**	P	*SN*	BI	73422	78822	78622	73822
377 423	**SN**	P	*SN*	BI	73423	78823	78623	73823
377 424	**SN**	P	*SN*	BI	73424	78824	78624	73824
377 425	**SN**	P	*SN*	BI	73425	78825	78625	73825
377 426	**SN**	P	*SN*	BI	73426	78826	78626	73826
377 427	**SN**	P	*SN*	BI	73427	78827	78627	73827
377 428	**SN**	P	*SN*	BI	73428	78828	78628	73828
377 429	**SN**	P	*SN*	BI	73429	78829	78629	73829
377 430	**SN**	P	*SN*	BI	73430	78830	78630	73830
377 431	**SN**	P	*SN*	BI	73431	78831	78631	73831
377 432	**SN**	P	*SN*	BI	73432	78832	78632	73832
377 433	**SN**	P	*SN*	BI	73433	78833	78633	73833
377 434	**SN**	P	*SN*	BI	73434	78834	78634	73834
377 435	**SN**	P	*SN*	BI	73435	78835	78635	73835
377 436	**SN**	P	*SN*	BI	73436	78836	78636	73836
377 437	**SN**	P	*SN*	BI	73437	78837	78637	73837
377 438	**SN**	P	*SN*	BI	73438	78838	78638	73838
377 439	**SN**	P	*SN*	BI	73439	78839	78639	73839
377 440	**SN**	P	*SN*	BI	73440	78840	78640	73840
377 441	**SN**	P	*SN*	BI	73441	78841	78641	73841
377 442	**SN**	P	*SN*	BI	73442	78842	78642	73842
377 443	**SN**	P	*SN*	BI	73443	78843	78643	73843
377 444	**SN**	P	*SN*	BI	73444	78844	78644	73844
377 445	**SN**	P	*SN*	BI	73445	78845	78645	73845
377 446	**SN**	P	*SN*	BI	73446	78846	78646	73846
377 447	**SN**	P	*SN*	BI	73447	78847	78647	73847
377 448	**SN**	P	*SN*	BI	73448	78848	78648	73848

377 449	**SN**	P	*SN*	BI	73449	78849	78649	73849
377 450	**SN**	P	*SN*	BI	73450	78850	78650	73850
377 451	**SN**	P	*SN*	BI	73451	78851	78651	73851
377 452	**SN**	P	*SN*	BI	73452	78852	78652	73852
377 453	**SN**	P	*SN*	BI	73453	78853	78653	73853
377 454	**SN**	P	*SN*	BI	73454	78854	78654	73854
377 455	**SN**	P	*SN*	BI	73455	78855	78655	73855
377 456	**SN**	P	*SN*	BI	73456	78856	78656	73856
377 457	**SN**	P	*SN*	BI	73457	78857	78657	73857
377 458	**SN**	P	*SN*	BI	73458	78858	78658	73858
377 459	**SN**	P	*SN*	BI	73459	78859	78659	73859
377 460	**SN**	P	*SN*	BI	73460	78860	78660	73860
377 461	**SN**	P	*SN*	BI	73461	78861	78661	73861
377 462	**SN**	P	*SN*	BI	73462	78862	78662	73862
377 463	**SN**	P	*SN*	BI	73463	78863	78663	73863
377 464	**SN**	P	*SN*	BI	73464	78864	78664	73864
377 465	**SN**	P	*SN*	BI	73465	78865	78665	73865
377 466	**SN**	P	*SN*	BI	73466	78866	78666	73866
377 467	**SN**	P	*SN*	BI	73467	78867	78667	73867
377 468	**SN**	P	*SN*	BI	73468	78868	78668	73868
377 469	**SN**	P	*SN*	BI	73469	78869	78669	73869
377 470	**SN**	P	*SN*	BI	73470	78870	78670	73870
377 471	**SN**	P	*SN*	BI	73471	78871	78671	73871
377 472	**SN**	P	*SN*	BI	73472	78872	78672	73872
377 473	**SN**	P	*SN*	BI	73473	78873	78673	73873
377 474	**SN**	P	*SN*	BI	73474	78874	78674	73874
377 475	**SN**	P	*SN*	BI	73475	78875	78675	73875

Class 377/5. 25 kV AC/750 V DC. DMCO–MSO–PTSO–DMCO. Dual voltage First Capital Connect units (sub-leased from Southern). Details as Class 377/2 unless stated.

DMCO(A). Bombardier Derby 2008–2009. 10/48. 43.1 t.
MSO. Bombardier Derby 2008–2009. –/69 1T. 39.3 t.
PTSO. Bombardier Derby 2008–2009. –/53 1TD 2W. 35.3 t.
DMCO(B). Bombardier Derby 2008–2009. 10/48. 43.1 t.

377 501	**FU**	P	*FC*	BF	73501	75901	74901	73601
377 502	**FU**	P	*FC*	BF	73502	75902	74902	73602
377 503	**FU**	P	*FC*	BF	73503	75903	74903	73603
377 504	**FU**	P	*FC*	BF	73504	75904	74904	73604
377 505	**FU**	P	*FC*	BF	73505	75905	74905	73605
377 506	**FU**	P	*FC*	BF	73506	75906	74906	73606
377 507	**FU**	P	*FC*	BF	73507	75907	74907	73607
377 508	**FU**	P	*FC*	BF	73508	75908	74908	73608
377 509	**FU**	P	*FC*	BF	73509	75909	74909	73609
377 510	**FU**	P	*FC*	BF	73510	75910	74910	73610
377 511	**FU**	P	*FC*	BF	73511	75911	74911	73611
377 512	**FU**	P	*FC*	BF	73512	75912	74912	73612
377 513	**FU**	P	*FC*	BF	73513	75913	74913	73613
377 514	**FU**	P	*FC*	BF	73514	75914	74914	73614
377 515	**FU**	P	*FC*	BF	73515	75915	74915	73615

377 516	**FU**	P	*FC*	BF	73516	75916	74916	73616
377 517	**FU**	P	*FC*	BF	73517	75917	74917	73617
377 518	**FU**	P	*FC*	BF	73518	75918	74918	73618
377 519	**FU**	P	*FC*	BF	73519	75919	74919	73619
377 520	**FU**	P	*FC*	BF	73520	75920	74920	73620
377 521	**FU**	P	*FC*	BF	73521	75921	74921	73621
377 522	**FU**	P	*FC*	BF	73522	75922	74922	73622
377 523	**FU**	P	*FC*	BF	73523	75923	74923	73623

CLASS 378 CAPITALSTAR BOMBARDIER DERBY

57 new Class 378 suburban Electrostars (designated Capitalstars by TfL) are being delivered for the London Overground network.

Formation: DMSO–MSO–TSO–DMSO or DMSO–MSO–PTSO–DMSO.
System: Class 378/1 750 V DC third rail only. Class 378/2 25 kV AC overhead and 750 V DC third rail.
Construction: Welded aluminium alloy underframe, sides and roof with steel ends. All sections bolted together.
Traction Motors: Two Bombardier asynchronous of 250 kW.
Wheel Arrangement: 2-Bo + 2-Bo + 2-2 + Bo-2.
Braking: Disc & regenerative.
Dimensions: 20.46/20.14 x 2.80 m.
Bogies: Bombardier P3-25/T3-25.
Couplers: Dellner 12.
Gangways: Within unit + end doors.
Control System: IGBT Inverter.
Doors: Sliding.
Maximum Speed: 75 m.p.h.
Heating & ventilation: Air conditioning.
Seating Layout: Longitudinal ("tube style") low density.
Multiple Working: Within class and with Classes 375, 376 and 377.

Class 378/1. 750 V DC. DMSO–MSO–TSO–DMSO. Third rail only units for East London Line services. Provision for retro-fitting as dual voltage if required.

DMSO(A). Bombardier Derby 2009–2010. –/36. 43.5 t.
MSO. Bombardier Derby 2009–2010. –/40. 39.4 t.
TSO. Bombardier Derby 2009–2010. –/34(6) 2W. 34.3 t.
DMSO(B). Bombardier Derby 2009–2010. –/36. 43.1 t.

Note: 378 150–154 fitted with de-icing equipment. TSO weighs 34.8 t.

378 135	**LO**	QW	*LO*	NG	38035	38235	38335	38135
378 136	**LO**	QW	*LO*	NG	38036	38236	38336	38136
378 137	**LO**	QW	*LO*	NG	38037	38237	38337	38137
378 138	**LO**	QW	*LO*	NG	38038	38238	38338	38138
378 139	**LO**	QW	*LO*	NG	38039	38239	38339	38139
378 140	**LO**	QW	*LO*	NG	38040	38240	38340	38140
378 141	**LO**	QW	*LO*	NG	38041	38241	38341	38141
378 142	**LO**	QW	*LO*	NG	38042	38242	38342	38142
378 143	**LO**	QW	*LO*	NG	38043	38243	38343	38143
378 144	**LO**	QW	*LO*	NG	38044	38244	38344	38144
378 145	**LO**	QW	*LO*	NG	38045	38245	38345	38145
378 146	**LO**	QW	*LO*	NG	38046	38246	38346	38146
378 147	**LO**	QW	*LO*	NG	38047	38247	38347	38147
378 148	**LO**	QW	*LO*	NG	38048	38248	38348	38148

378 149	**LO**	QW	*LO*	NG	38049	38249	38349	38149
378 150	**LO**	QW	*LO*	NG	38050	38250	38350	38150
378 151	**LO**	QW	*LO*	NG	38051	38251	38351	38151
378 152	**LO**	QW	*LO*	NG	38052	38252	38352	38152
378 153	**LO**	QW	*LO*	NG	38053	38253	38353	38153
378 154	**LO**	QW	*LO*	NG	38054	38254	38354	38154

Class 378/2. 25 kV AC/750 V DC. DMSO–MSO–PTSO–DMSO. Dual voltage units mainly used on North London Railway services. 378 201–224 built as 378 001–024 (3-car units). These sets are returning to Derby for fitting with an extra MSO vehicle (382xx series). Vehicles still to be inserted at time of going to press are shown in *italics* (all units should be 4-cars by late 2010).

DMSO(A). Bombardier Derby 2008–2010. –/36. 43.2 t.
MSO. Bombardier Derby 2008–2010. –/40. 39.8 t.
PTSO. Bombardier Derby 2008–2010. –/34(6) 2W. 39.0 t.
DMSO(B). Bombardier Derby 2008–2010. –/36. 42.8 t.

378 201	**LO**	QW	*LO*	NG	38001	38201	38301	38101
378 202	**LO**	QW	*LO*	NG	38002	38202	38302	38102
378 203	**LO**	QW	*LO*	NG	38003	*38203*	38303	38103
378 204	**LO**	QW	*LO*	NG	38004	*38204*	38304	38104
378 205	**LO**	QW	*LO*	NG	38005	*38205*	38305	38105
378 206	**LO**	QW	*LO*	NG	38006	*38206*	38306	38106
378 207	**LO**	QW	*LO*	NG	38007	*38207*	38307	38107
378 208	**LO**	QW	*LO*	NG	38008	38208	38308	38108
378 209	**LO**	QW	*LO*	NG	38009	*38209*	38309	38109
378 210	**LO**	QW	*LO*	NG	38010	38210	38310	38110
378 211	**LO**	QW	*LO*	NG	38011	38211	38311	38111
378 212	**LO**	QW	*LO*	NG	38012	*38212*	38312	38112
378 213	**LO**	QW	*LO*	NG	38013	*38213*	38313	38113
378 214	**LO**	QW	*LO*	NG	38014	38214	38314	38114
378 215	**LO**	QW	*LO*	NG	38015	*38215*	38315	38115
378 216	**LO**	QW	*LO*	NG	38016	38216	38316	38116
378 217	**LO**	QW	*LO*	NG	38017	*38217*	38317	38117
378 218	**LO**	QW	*LO*	NG	38018	*38218*	38318	38118
378 219	**LO**	QW	*LO*	NG	38019	*38219*	38319	38119
378 220	**LO**	QW	*LO*	NG	38020	*38220*	38320	38120
378 221	**LO**	QW	*LO*	NG	38021	*38221*	38321	38121
378 222	**LO**	QW	*LO*	NG	38022	38222	38322	38122
378 223	**LO**	QW	*LO*	NG	38023	*38223*	38323	38123
378 224	**LO**	QW	*LO*	NG	38024	*38224*	38324	38124

378 225	**LO**	QW	*LO*	NG	38025	38225	38325	38125
378 226	**LO**	QW	*LO*	NG	38026	38226	38326	38126
378 227	**LO**	QW	*LO*	NG	38027	38227	38327	38127
378 228	**LO**	QW	*LO*	NG	38028	38228	38328	38128
378 229	**LO**	QW	*LO*	NG	38029	38229	38329	38129
378 230	**LO**	QW	*LO*	NG	38030	38230	38330	38130
378 231	**LO**	QW	*LO*	NG	38031	38231	38331	38131
378 232	**LO**	QW	*LO*	NG	38032	38232	38332	38132
378 233	**LO**	QW	*LO*	NG	38033	38233	38333	38133
378 234	**LO**	QW	*LO*	NG	38034	38234	38334	38134

378 255	**LO**	QW	38055	38255	38355	38155
378 256	**LO**	QW	38056	38256	38356	38156
378 257	**LO**	QW	38057	38257	38357	38157

CLASS 379 ELECTROSTAR BOMBARDIER DERBY

30 new 4-car Class 379 Bombardier EMUs are on order for National Express East Anglia (principally London Liverpool Street–Stansted Airport services), with the first due to be delivered in late 2010. Full details awaited.

Formation: DMSO–MSO–PTSO–DMCO.
System: 25 kV AC overhead.
Construction: Welded aluminium alloy underframe, sides and roof with steel ends. All sections bolted together.
Traction Motors: Two Bombardier asynchronous of 250 kW.
Wheel Arrangement: 2-Bo + 2-Bo + 2-2 + Bo-2.
Braking: Disc & regenerative.
Dimensions: 20.0 x 2.80 m.
Bogies: Bombardier P3-25/T3-25.
Couplers: Dellner 12.
Gangways: Throughout.
Control System: IGBT Inverter.
Doors: Sliding plug.
Maximum Speed: 100 m.p.h.
Heating & ventilation: Air conditioning.
Seating Layout: 1: 2+1 facing. 2: 2+2 facing/unidirectional.
Multiple Working: Within class and with Classes 375, 376 and 378.

DMSO. Bombardier Derby 2010–2011. . . t.
MSO. Bombardier Derby 2010–2011. . . t.
PTSO. Bombardier Derby 2010–2011. . . t.
DMCO. Bombardier Derby 2010–2011. . . t.

379 001	LY	61201	61701	61901	62101
379 002	LY	61202	61702	61902	62102
379 003	LY	61203	61703	61903	62103
379 004	LY	61204	61704	61904	62104
379 005	LY	61205	61705	61905	62105
379 006	LY	61206	61706	61906	62106
379 007	LY	61207	61707	61907	62107
379 008	LY	61208	61708	61908	62108
379 009	LY	61209	61709	61909	62109
379 010	LY	61210	61710	61910	62110
379 011	LY	61211	61711	61911	62111
379 012	LY	61212	61712	61912	62112
379 013	LY	61213	61713	61913	62113
379 014	LY	61214	61714	61914	62114
379 015	LY	61215	61715	61915	62115
379 016	LY	61216	61716	61916	62116
379 017	LY	61217	61717	61917	62117
379 018	LY	61218	61718	61918	62118
379 019	LY	61219	61719	61919	62119
379 020	LY	61220	61720	61920	62120
379 021	LY	61221	61721	61921	62121
379 022	LY	61222	61722	61922	62122
379 023	LY	61223	61723	61923	62123

379 024	LY	61224	61724	61924	62124
379 025	LY	61225	61725	61925	62125
379 026	LY	61226	61726	61926	62126
379 027	LY	61227	61727	61927	62127
379 028	LY	61228	61728	61928	62128
379 029	LY	61229	61729	61929	62129
379 030	LY	61230	61730	61930	62130

CLASS 380 DESIRO UK SIEMENS

38 new Class 380 Siemens EMUs are currently being delivered for ScotRail, mainly for Strathclyde area services.

Formation: DMSO–PTSO–DMSO or DMSO–PTSO–TSO–DMSO.
System: 25 kV AC overhead.
Construction: Welded aluminium with steel ends.
Traction Motors: Four Siemens ITB2016-0GB02 asynchronous of 250 kW.
Wheel Arrangement: Bo-Bo + 2-2 (+2-2) + Bo-Bo
Braking: Disc & regenerative.
Dimensions: 23.36/23.78 x 2.80 m.
Bogies: SGP SF5000.
Couplers: Voith.
Gangways: Throughout.
Control System: IGBT Inverter.
Doors: Sliding plug.
Maximum Speed: 100 m.p.h.
Heating & ventilation: Air conditioning.
Seating Layout: 2+2 facing/unidirectional.
Multiple Working: Within class.

DMSO(A). Siemens Uerdingen (Krefeld) 2009–2010. –/70. 45.1 t.
PTSO. Siemens Uerdingen (Krefeld) 2009–2010. –/57(12) 1TD 2W. 42.4 t.
TSO. Siemens Uerdingen (Krefeld) 2009–2010. –/74 1T. 34.7 t.
DMSO(B). Siemens Uerdingen (Krefeld) 2009–2010. –/64(5). 45.3 t.

Class 380/0. 3-car units.

380 001	**SR**	E	38501	38601		38701
380 002	**SR**	E	38502	38602		38702
380 003	**SR**	E	38503	38603		38703
380 004	**SR**	E	38504	38604		38704
380 005	**SR**	E	38505	38605		38705
380 006	**SR**	E	38506	38606		38706
380 007	**SR**	E	38507	38607		38707
380 008	**SR**	E	38508	38608		38708
380 009	**SR**	E	38509	38609		38709
380 010	**SR**	E	38510	38610		38710
380 011	**SR**	E	38511	38611		38711
380 012	**SR**	E	38512	38612		38712
380 013	**SR**	E	38513	38613		38713
380 014	**SR**	E	38514	38614		38714
380 015	**SR**	E	38515	38615		38715
380 016	**SR**	E	38516	38616		38716
380 017	**SR**	E	38517	38617		38717
380 018	**SR**	E	38518	38618		38718
380 019	**SR**	E	38519	38619		38719

380 020	**SR**	E		38520	38620		38720
380 021	**SR**	E		38521	38621		38721
380 022	**SR**	E		38522	38622		38722

Class 380/1. 4-car units.

380 101	**SR**	E		38551	38651	38851	38751
380 102	**SR**	E		38552	38652	38852	38752
380 103	**SR**	E		38553	38653	38853	38753
380 104	**SR**	E		38554	38654	38854	38754
380 105	**SR**	E	GW	38555	38655	38855	38755
380 106	**SR**	E	GW	38556	38656	38856	38756
380 107	**SR**	E	GW	38557	38657	38857	38757
380 108	**SR**	E	GW	38558	38658	38858	38758
380 109	**SR**	E	GW	38559	38659	38859	38759
380 110	**SR**	E	GW	38560	38660	38860	38760
380 111	**SR**	E		38561	38661	38861	38761
380 112	**SR**	E		38562	38662	38862	38762
380 113	**SR**	E		38563	38663	38863	38763
380 114	**SR**	E		38564	38664	38864	38764
380 115	**SR**	E		38565	38665	38865	38765
380 116	**SR**	E		38566	38666	38866	38766

CLASS 390 PENDOLINO ALSTOM

Tilting West Coast Main Line units.

Formation: DMRFO–MFO–PTFO–MFO–(TSO)–(MSO)–TSO–MSO–PTSRMB–MSO–DMSO.
Construction: Welded aluminium alloy.
Traction Motors: Two Alstom ONIX 800 of 425 kW.
Wheel Arrangement: 1A-A1 + 1A-A1 + 2-2 + 1A-A1 (+ 2-2 + 1A-A1) + 2-2 + 1A-A1 + 2-2 + 1A-A1 + 1A-A1.
Braking: Disc, rheostatic & regenerative.
Dimensions: 24.80/23.90 x 2.73 m.
Bogies: Fiat-SIG.
Couplers: Dellner 12.
Gangways: Within unit.
Control System: IGBT Inverter.
Doors: Sliding plug.
Maximum Speed: 125 m.p.h.
Heating & ventilation: Air conditioning.
Seating Layout: 1: 2+1 facing/unidirectional, 2: 2+2 facing/unidirectional.
Multiple Working: Within class. Can also be controlled from Class 57/3 locos.

DMRFO: Alstom Birmingham 2001–2005. 18/–. 55.6 t.
MFO(A): Alstom Birmingham 2001–2005. 37/–(2) 1TD 1W. 52.0 t.
PTFO: Alstom Birmingham 2001–2005. 44/– 1T. 50.1 t.
MFO(B): Alstom Birmingham 2001–2005. 46/– 1T. 51.8 t.
(**TSO**: Alstom Savigliano 2010–2011. –/74 1T. . t.)
(**MSO**: Alstom Savigliano 2010–2011. –/76 1T. . t.)
TSO: Alstom Birmingham 2001–2005. –/76 1T. 45.5 t.
MSO(A): Alstom Birmingham 2001–2005. –/62(4) 1TD 1W. 50.0 t.
PTSRMB: Alstom Birmingham 2001–2005. –/48. 52.0 t.
MSO(B): Alstom Birmingham 2001–2005. –/62(2) 1TD 1W. 51.7 t.
DMSO: Alstom Birmingham 2001–2005. –/46 1T. 51.0 t.

Advertising livery: 390 004 – Black Alstom vinyls on Virgin silver livery.

Notes: Units up to 390 034 were delivered as 8-car sets, without the TSO (688xx). During 2004–05 these units had their 9th cars added.

62 extra vehicles are on order to lengthen 31 sets to 11-cars, but it has not yet been finalised which sets these extra vehicles will be added to. Four new complete 11-car units (390 054–057) are also under construction.

390 033 was written off following accident damage in the Lambrigg accident of February 2007.

390 001	**VT**	A	*VW*	MA	69101	69401	69501	69601	68801
					69701	69801	69901	69201	
390 002	**VT**	A	*VW*	MA	69102	69402	69502	69602	68802
					69702	69802	69902	69202	
390 003	**VT**	A	*VW*	MA	69103	69403	69503	69603	68803
					69703	69803	69903	69203	
390 004	**AL**	A	*VW*	MA	69104	69404	69504	69604	68804
					69704	69804	69904	69204	
390 005	**VT**	A	*VW*	MA	69105	69405	69505	69605	68805
					69705	69805	69905	69205	
390 006	**VT**	A	*VW*	MA	69106	69406	69506	69606	68806
					69706	69806	69906	69206	
390 007	**VT**	A	*VW*	MA	69107	69407	69507	69607	68807
					69707	69807	69907	69207	
390 008	**VT**	A	*VW*	MA	69108	69408	69508	69608	68808
					69708	69808	69908	69208	
390 009	**VT**	A	*VW*	MA	69109	69409	69509	69609	68809
					69709	69809	69909	69209	
390 010	**VT**	A	*VW*	MA	69110	69410	69510	69610	68810
					69710	69810	69910	69210	
390 011	**VT**	A	*VW*	MA	69111	69411	69511	69611	68811
					69711	69811	69911	69211	
390 012	**VT**	A	*VW*	MA	69112	69412	69512	69612	68812
					69712	69812	69912	69212	
390 013	**VT**	A	*VW*	MA	69113	69413	69513	69613	68813
					69713	69813	69913	69213	
390 014	**VT**	A	*VW*	MA	69114	69414	69514	69614	68814
					69714	69814	69914	69214	
390 015	**VT**	A	*VW*	MA	69115	69415	69515	69615	68815
					69715	69815	69915	69215	
390 016	**VT**	A	*VW*	MA	69116	69416	69516	69616	68816
					69716	69816	69916	69216	
390 017	**VT**	A	*VW*	MA	69117	69417	69517	69617	68817
					69717	69817	69917	69217	
390 018	**VT**	A	*VW*	MA	69118	69418	69518	69618	68818
					69718	69818	69918	69218	
390 019	**VT**	A	*VW*	MA	69119	69419	69519	69619	68819
					69719	69819	69919	69219	
390 020	**VT**	A	*VW*	MA	69120	69420	69520	69620	68820
					69720	69820	69920	69220	

390 021	**VT**	A	*VW*	MA	69121	69421	69521	69621	68821
					69721	69821	69921	69221	
390 022	**VT**	A	*VW*	MA	69122	69422	69522	69622	68822
					69722	69822	69922	69222	
390 023	**VT**	A	*VW*	MA	69123	69423	69523	69623	68823
					69723	69823	69923	69223	
390 024	**VT**	A	*VW*	MA	69124	69424	69524	69624	68824
					69724	69824	69924	69224	
390 025	**VT**	A	*VW*	MA	69125	69425	69525	69625	68825
					69725	69825	69925	69225	
390 026	**VT**	A	*VW*	MA	69126	69426	69526	69626	68826
					69726	69826	69926	69226	
390 027	**VT**	A	*VW*	MA	69127	69427	69527	69627	68827
					69727	69827	69927	69227	
390 028	**VT**	A	*VW*	MA	69128	69428	69528	69628	68828
					69728	69828	69928	69228	
390 029	**VT**	A	*VW*	MA	69129	69429	69529	69629	68829
					69729	69829	69929	69229	
390 030	**VT**	A	*VW*	MA	69130	69430	69530	69630	68830
					69730	69830	69930	69230	
390 031	**VT**	A	*VW*	MA	69131	69431	69531	69631	68831
					69731	69831	69931	69231	
390 032	**VT**	A	*VW*	MA	69132	69432	69532	69632	68832
					69732	69832	69932	69232	
390 034	**VT**	A	*VW*	MA	69134	69434	69534	69634	68834
					69734	69834	69934	69234	
390 035	**VT**	A	*VW*	MA	69135	69435	69535	69635	68835
					69735	69835	69935	69235	
390 036	**VT**	A	*VW*	MA	69136	69436	69536	69636	68836
					69736	69836	69936	69236	
390 037	**VT**	A	*VW*	MA	69137	69437	69537	69637	68837
					69737	69837	69937	69237	
390 038	**VT**	A	*VW*	MA	69138	69438	69538	69638	68838
					69738	69838	69938	69238	
390 039	**VT**	A	*VW*	MA	69139	69439	69539	69639	68839
					69739	69839	69939	69239	
390 040	**VT**	A	*VW*	MA	69140	69440	69540	69640	68840
					69740	69840	69940	69240	
390 041	**VT**	A	*VW*	MA	69141	69441	69541	69641	68841
					69741	69841	69941	69241	
390 042	**VT**	A	*VW*	MA	69142	69442	69542	69642	68842
					69742	69842	69942	69242	
390 043	**VT**	A	*VW*	MA	69143	69443	69543	69643	68843
					69743	69843	69943	69243	
390 044	**VT**	A	*VW*	MA	69144	69444	69544	69644	68844
					69744	69844	69944	69244	
390 045	**VT**	A	*VW*	MA	69145	69445	69545	69645	68845
					69745	69845	69945	69245	
390 046	**VT**	A	*VW*	MA	69146	69446	69546	69646	68846
					69746	69846	69946	69246	

390 047	**VT**	A	*VW*	MA	69147 69447 69547 69647 68847 69747 69847 69947 69247
390 048	**VT**	A	*VW*	MA	69148 69448 69548 69648 68848 69748 69848 69948 69248
390 049	**VT**	A	*VW*	MA	69149 69449 69549 69649 68849 69749 69849 69949 69249
390 050	**VT**	A	*VW*	MA	69150 69450 69550 69650 68850 69750 69850 69950 69250
390 051	**VT**	A	*VW*	MA	69151 69451 69551 69651 68851 69751 69851 69951 69251
390 052	**VT**	A	*VW*	MA	69152 69452 69552 69652 68852 69752 69852 69952 69252
390 053	**VT**	A	*VW*	MA	69153 69453 69553 69653 68853 69753 69853 69953 69253
390 054	**VT**	A			69154 69454 69554 69654 65354 68954 68854 69754 69854 69954 69254
390 055	**VT**	A			69155 69455 69555 69655 65355 68955 68855 69755 69855 69955 69255
390 056	**VT**	A			69156 69456 69556 69656 65356 68956 68856 69756 69856 69956 69256
390 057	**VT**	A			69157 69457 69557 69657 65357 68957 68857 69757 69857 69957 69257

Names (carried on MFO No. 696xx):

390 001	Virgin Pioneer
390 002	Virgin Angel
390 003	Virgin Hero
390 004	Alstom Pendolino
390 005	City of Wolverhampton
390 006	Tate Liverpool
390 007	Virgin Lady
390 008	Virgin King
390 009	Treaty of Union
390 010	A Decade of Progress
390 011	City of Lichfield
390 012	Virgin Star
390 013	Virgin Spirit
390 014	City of Manchester
390 015	Virgin Crusader
390 016	Virgin Champion
390 017	Virgin Prince
390 018	Virgin Princess
390 019	Virgin Warrior
390 020	Virgin Cavalier
390 021	Virgin Dream
390 022	Penny the Pendolino
390 023	Virgin Glory
390 024	Virgin Venturer
390 025	Virgin Stagecoach
390 026	Virgin Enterprise
390 027	Virgin Buccaneer
390 028	City of Preston
390 029	City of Stoke-on-Trent
390 030	City of Edinburgh
390 031	City of Liverpool
390 032	City of Birmingham
390 034	City of Carlisle
390 035	City of Lancaster
390 036	City of Coventry
390 037	Virgin Difference
390 038	City of London
390 039	Virgin Quest
390 040	Virgin Pathfinder
390 041	City of Chester
390 042	City of Bangor/Dinas Bangor
390 043	Virgin Explorer
390 044	Virgin Lionheart
390 045	101 Squadron
390 046	Virgin Soldiers
390 047	CLIC Sargent
390 048	Virgin Harrier
390 049	Virgin Express
390 050	Virgin Invader
390 051	Virgin Ambassador
390 052	Virgin Knight
390 053	Mission Accomplished

CLASS 395 HS1 DOMESTIC SETS HITACHI JAPAN

New 6-car dual-voltage units for Southeastern domestic services from St Pancras International to Ashford/Dover/Margate via Ramsgate and Faversham.

Formation: PDTSO–MSO–MSO–MSO–MSO–PDTSO.
Systems: 25 kV AC overhead/750 V DC third rail.
Construction: Aluminium.
Traction Motors: Hitachi asynchronous of 210 kW.
Wheel Arrangement: 2-2 + Bo-Bo + Bo-Bo + Bo-Bo + Bo-Bo + 2-2.
Braking: Disc, rheostatic & capability for regenerative braking.
Dimensions: 20.88/20.0 x 2.81 m.
Couplers: Scharfenberg.
Bogies: Hitachi.
Control System: IGBT Inverter.
Gangways: Within unit.
Maximum Speed: 140 m.p.h.
Doors: Single-leaf sliding.
Multiple Working: Within class only.
Heating & ventilation: Air conditioning.
Seating Layout: 2+2 facing/unidirectional (mainly unidirectional).

PDTSO(A): Hitachi Kasado, Japan 2006–2009. –/28(12) 1TD 2W. 46.7 t.
MSO: Hitachi Kasado, Japan 2006–2009. –/66. 45.0 t. 45.7 t.
PDTSO(B): Hitachi Kasado, Japan 2006–2009. –/48 1T. 46.7 t.

395 001	**SB**	E	*SE*	AD	39011	39012	39013	39014	39015	39016
395 002	**SB**	E	*SE*	AD	39021	39022	39023	39024	39025	39026
395 003	**SB**	E	*SE*	AD	39031	39032	39033	39034	39035	39036
395 004	**SB**	E	*SE*	AD	39041	39042	39043	39044	39045	39046
395 005	**SB**	E	*SE*	AD	39051	39052	39053	39054	39055	39056
395 006	**SB**	E	*SE*	AD	39061	39062	39063	39064	39065	39066
395 007	**SB**	E	*SE*	AD	39071	39072	39073	39074	39075	39076
395 008	**SB**	E	*SE*	AD	39081	39082	39083	39084	39085	39086
395 009	**SB**	E	*SE*	AD	39091	39092	39093	39094	39095	39096
395 010	**SB**	E	*SE*	AD	39101	39102	39103	39104	39105	39106
395 011	**SB**	E	*SE*	AD	39111	39112	39113	39114	39115	39116
395 012	**SB**	E	*SE*	AD	39121	39122	39123	39124	39125	39126
395 013	**SB**	E	*SE*	AD	39131	39132	39133	39134	39135	39136
395 014	**SB**	E	*SE*	AD	39141	39142	39143	39144	39145	39146
395 015	**SB**	E	*SE*	AD	39151	39152	39153	39154	39155	39156
395 016	**SB**	E	*SE*	AD	39161	39162	39163	39164	39165	39166
395 017	**SB**	E	*SE*	AD	39171	39172	39173	39174	39175	39176
395 018	**SB**	E	*SE*	AD	39181	39182	39183	39184	39185	39186
395 019	**SB**	E	*SE*	AD	39191	39192	39193	39194	39195	39196
395 020	**SB**	E	*SE*	AD	39201	39202	39203	39204	39205	39206
395 021	**SB**	E	*SE*	AD	39211	39212	39213	39214	39215	39216
395 022	**SB**	E	*SE*	AD	39221	39222	39223	39224	39225	39226
395 023	**SB**	E	*SE*	AD	39231	39232	39233	39234	39235	39236
395 024	**SB**	E	*SE*	AD	39241	39242	39243	39244	39245	39246
395 025	**SB**	E	*SE*	AD	39251	39252	39253	39254	39255	39256
395 026	**SB**	E	*SE*	AD	39261	39262	39263	39264	39265	39266
395 027	**SB**	E	*SE*	AD	39271	39272	39273	39274	39275	39276
395 028	**SB**	E	*SE*	AD	39281	39282	39283	39284	39285	39286
395 029	**SB**	E	*SE*	AD	39291	39292	39293	39294	39295	39296

Names (carried on end cars):

395 001	Dame Kelly Holmes	395 006	Daley Thompson
395 002	Sebastian Coe	395 007	Steve Backley
395 003	Sir Steve Redgrave	395 008	Ben Ainslie
395 004	Sir Chris Hoy	395 009	Rebecca Adlington
395 005	Dame Tanni Grey-Thompson	395 016	Jamie Staff

2. 750 V DC THIRD RAIL EMUs

These classes use the third rail system at 750 V DC (unless stated). Outer couplers are buckeyes on units built before 1982 with bar couplers within the units. Newer units generally have Dellner outer couplers.

CLASS 442 WESSEX EXPRESS BREL DERBY

Stock built for Waterloo–Bournemouth–Weymouth services. Withdrawn from service with South West Trains in early 2007. All units have now been refurbished at Railcare, Wolverton for use by Southern principally on Victoria–Gatwick Airport–Brighton services.

Formation: DTSO(A)–TSO–MBC–TSO(W)–DTSO(B).
Construction: Steel.
Traction Motors: Four EE546 of 300 kW recovered from Class 432s.
Wheel Arrangement: 2-2 + 2-2 + Bo-Bo + 2-2 + 2-2.
Braking: Disc.
Dimensions: 23.15/23.00 x 2.74 m.
Bogies: Two BREL P7 motor bogies (MBSO). T4 bogies (trailer cars).
Couplers: Buckeye.
Control System: 1986-type.
Gangways: Throughout.
Maximum Speed: 100 m.p.h.
Doors: Sliding plug.
Heating & Ventilation: Air conditioning.
Seating Layout: 1: 2+1 facing, 2: 2+2 mainly unidirectional.
Multiple Working: Within class and with locos of Classes 33/1 & 73 in an emergency.

DTSO(A). Lot No. 31030 Derby 1988–1989. –/74. 38.5 t.
TSO. Lot No. 31032 Derby 1988–1989. –/76 2T. 37.5 t.
MBC. Lot No. 31034 Derby 1988–1989. 24/28. 55.0 t.
TSO(W). Lot No. 31033 Derby 1988–1989. –/66(4) 1TD 1T 2W. 37.8 t.
DTSO(B). Lot No. 31031 Derby 1988–1989. –/74. 37.3 t.

442 401	**GV**	A	*SN*	BI	77382	71818	62937	71842	77406
442 402	**GV**	A	*SN*	BI	77383	71819	62938	71843	77407
442 403	**GV**	A	*SN*	BI	77384	71820	62941	71844	77408
442 404	**GV**	A	*SN*	BI	77385	71821	62939	71845	77409
442 405	**GV**	A	*SN*	BI	77386	71822	62944	71846	77410
442 406	**GV**	A	*SN*	BI	77389	71823	62942	71847	77411
442 407	**GV**	A	*SN*	BI	77388	71824	62943	71848	77412
442 408	**GV**	A	*SN*	BI	77387	71825	62945	71849	77413
442 409	**GV**	A	*SN*	BI	77390	71826	62946	71850	77414

442 410	**GV**	A	*SN*	BI	77391	71827	62948	71851	77415
442 411	**GV**	A	*SN*	BI	77392	71828	62940	71858	77422
442 412	**GV**	A	*SN*	SI	77393	71829	62947	71853	77417
442 413	**GV**	A	*SN*	BI	77394	71830	62949	71854	77418
442 414	**GV**	A	*SN*	BI	77395	71831	62950	71855	77419
442 415	**GV**	A	*SN*	BI	77396	71832	62951	71856	77420
442 416	**GV**	A	*SN*	BI	77397	71833	62952	71857	77421
442 417	**GV**	A	*SN*	BI	77398	71834	62953	71852	77416
442 418	**GV**	A	*SN*	BI	77399	71835	62954	71859	77423
442 419	**GV**	A	*SN*	BI	77400	71836	62955	71860	77424
442 420	**GV**	A	*SN*	BI	77401	71837	62956	71861	77425
442 421	**GV**	A	*SN*	BI	77402	71838	62957	71862	77426
442 422	**GV**	A	*SN*	BI	77403	71839	62958	71863	77427
442 423	**GV**	A	*SN*	BI	77404	71840	62959	71864	77428
442 424	**GV**	A	*SN*	BI	77405	71841	62960	71865	77429

CLASS 444 DESIRO UK SIEMENS

Express units.

Formation: DMCO–TSO–TSO–TSORMB–DMSO.
Construction: Aluminium.
Traction Motors: 4 Siemens 1TB2016-0GB02 asynchronous of 250 kW.
Wheel Arrangement: Bo-Bo + 2-2 + 2-2 + 2-2 + Bo-Bo.
Braking: Disc, rheostatic & regenerative.
Dimensions: 23.57 x 2.80 m.
Bogies: SGP SF5000.
Couplers: Dellner 12.
Gangways: Throughout.
Control System: IGBT Inverter.
Doors: Single-leaf sliding plug.
Maximum Speed: 100 m.p.h.
Heating & Ventilation: Air conditioning.
Seating Layout: 1: 2+1 facing/unidirectional, 2: 2+2 facing/unidirectional.
Multiple Working: Within class and with Class 450.

DMSO. Siemens Vienna/Krefeld 2003–2004. –/76. 51.3 t.
TSO 67101–67145. Siemens Vienna/Krefeld 2003–2004. –/76 1T. 40.3 t.
TSO 67151–67195. Siemens Vienna/Krefeld 2003–2004. –/76 1T. 36.8 t.
TSORMB. Siemens Vienna/Krefeld 2003–2004. –/47 1T 1TD 2W. 42.1 t.
DMCO. Siemens Vienna/Krefeld 2003–2004. 35/24. 51.3 t.

444 001	**ST**	A	*SW*	NT	63801	67101	67151	67201	63851
444 002	**ST**	A	*SW*	NT	63802	67102	67152	67202	63852
444 003	**ST**	A	*SW*	NT	63803	67103	67153	67203	63853
444 004	**ST**	A	*SW*	NT	63804	67104	67154	67204	63854
444 005	**ST**	A	*SW*	NT	63805	67105	67155	67205	63855
444 006	**ST**	A	*SW*	NT	63806	67106	67156	67206	63856
444 007	**ST**	A	*SW*	NT	63807	67107	67157	67207	63857
444 008	**ST**	A	*SW*	NT	63808	67108	67158	67208	63858
444 009	**ST**	A	*SW*	NT	63809	67109	67159	67209	63859
444 010	**ST**	A	*SW*	NT	63810	67110	67160	67210	63860
444 011	**ST**	A	*SW*	NT	63811	67111	67161	67211	63861
444 012	**ST**	A	*SW*	NT	63812	67112	67162	67212	63862
444 013	**ST**	A	*SW*	NT	63813	67113	67163	67213	63863
444 014	**ST**	A	*SW*	NT	63814	67114	67164	67214	63864

444 015	**ST**	A	*SW*	NT	63815	67115	67165	67215	63865
444 016	**ST**	A	*SW*	NT	63816	67116	67166	67216	63866
444 017	**ST**	A	*SW*	NT	63817	67117	67167	67217	63867
444 018	**ST**	A	*SW*	NT	63818	67118	67168	67218	63868
444 019	**ST**	A	*SW*	NT	63819	67119	67169	67219	63869
444 020	**ST**	A	*SW*	NT	63820	67120	67170	67220	63870
444 021	**ST**	A	*SW*	NT	63821	67121	67171	67221	63871
444 022	**ST**	A	*SW*	NT	63822	67122	67172	67222	63872
444 023	**ST**	A	*SW*	NT	63823	67123	67173	67223	63873
444 024	**ST**	A	*SW*	NT	63824	67124	67174	67224	63874
444 025	**ST**	A	*SW*	NT	63825	67125	67175	67225	63875
444 026	**ST**	A	*SW*	NT	63826	67126	67176	67226	63876
444 027	**ST**	A	*SW*	NT	63827	67127	67177	67227	63877
444 028	**ST**	A	*SW*	NT	63828	67128	67178	67228	63878
444 029	**ST**	A	*SW*	NT	63829	67129	67179	67229	63879
444 030	**ST**	A	*SW*	NT	63830	67130	67180	67230	63880
444 031	**ST**	A	*SW*	NT	63831	67131	67181	67231	63881
444 032	**ST**	A	*SW*	NT	63832	67132	67182	67232	63882
444 033	**ST**	A	*SW*	NT	63833	67133	67183	67233	63883
444 034	**ST**	A	*SW*	NT	63834	67134	67184	67234	63884
444 035	**ST**	A	*SW*	NT	63835	67135	67185	67235	63885
444 036	**ST**	A	*SW*	NT	63836	67136	67186	67236	63886
444 037	**ST**	A	*SW*	NT	63837	67137	67187	67237	63887
444 038	**ST**	A	*SW*	NT	63838	67138	67188	67238	63888
444 039	**ST**	A	*SW*	NT	63839	67139	67189	67239	63889
444 040	**ST**	A	*SW*	NT	63840	67140	67190	67240	63890
444 041	**ST**	A	*SW*	NT	63841	67141	67191	67241	63891
444 042	**ST**	A	*SW*	NT	63842	67142	67192	67242	63892
444 043	**ST**	A	*SW*	NT	63843	67143	67193	67243	63893
444 044	**ST**	A	*SW*	NT	63844	67144	67194	67244	63894
444 045	**ST**	A	*SW*	NT	63845	67145	67195	67245	63895

Names (carried on TSORMB):

444 001 NAOMI HOUSE
444 012 DESTINATION WEYMOUTH
444 018 THE FAB 444

CLASS 450 DESIRO UK SIEMENS

Outer suburban units.

Formation: DMSO–TCO–TSO–DMSO (DMSO–TSO–TCO–DMSO 450 111–127).
Construction: Aluminium.
Traction Motors: 4 Siemens 1TB2016-0GB02 asynchronous of 250 kW.
Wheel Arrangement: Bo-Bo + 2-2 + 2-2 + Bo-Bo.
Braking: Disc, rheostatic & regenerative.
Bogies: SGP SF5000.
Gangways: Throughout.
Doors: Sliding plug.
Heating & Ventilation: Air conditioning.
Seating Layout: 1: 2+2 facing/unidirectional, 2: 3+2 facing/unidirectional.
Dimensions: 20.34 x 2.79 m.
Couplers: Dellner 12.
Control System: IGBT Inverter.
Maximum Speed: 100 m.p.h.

Multiple Working: Within class and with Class 444.

Class 450/0. Standard units.

DMSO(A). Siemens Krefeld/Vienna 2002–2006. –/70. 48.0 t.
TCO. Siemens Krefeld/Vienna 2002–2006. 24/32(4) 1T. 35.8 t.
TSO. Siemens Krefeld/Vienna 2002–2006. –/61(9) 1TD 2W. 39.8 t.
DMSO(B). Siemens Krefeld/Vienna 2002–2006. –/70. 48.6 t.

450 001	**SD**	A	*SW*	NT	63201	64201	68101	63601
450 002	**SD**	A	*SW*	NT	63202	64202	68102	63602
450 003	**SD**	A	*SW*	NT	63203	64203	68103	63603
450 004	**SD**	A	*SW*	NT	63204	64204	68104	63604
450 005	**SD**	A	*SW*	NT	63205	64205	68105	63605
450 006	**SD**	A	*SW*	NT	63206	64206	68106	63606
450 007	**SD**	A	*SW*	NT	63207	64207	68107	63607
450 008	**SD**	A	*SW*	NT	63208	64208	68108	63608
450 009	**SD**	A	*SW*	NT	63209	64209	68109	63609
450 010	**SD**	A	*SW*	NT	63210	64210	68110	63610
450 011	**SD**	A	*SW*	NT	63211	64211	68111	63611
450 012	**SD**	A	*SW*	NT	63212	64212	68112	63612
450 013	**SD**	A	*SW*	NT	63213	64213	68113	63613
450 014	**SD**	A	*SW*	NT	63214	64214	68114	63614
450 015	**SD**	A	*SW*	NT	63215	64215	68115	63615
450 016	**SD**	A	*SW*	NT	63216	64216	68116	63616
450 017	**SD**	A	*SW*	NT	63217	64217	68117	63617
450 018	**SD**	A	*SW*	NT	63218	64218	68118	63618
450 019	**SD**	A	*SW*	NT	63219	64219	68119	63619
450 020	**SD**	A	*SW*	NT	63220	64220	68120	63620
450 021	**SD**	A	*SW*	NT	63221	64221	68121	63621
450 022	**SD**	A	*SW*	NT	63222	64222	68122	63622
450 023	**SD**	A	*SW*	NT	63223	64223	68123	63623
450 024	**SD**	A	*SW*	NT	63224	64224	68124	63624
450 025	**SD**	A	*SW*	NT	63225	64225	68125	63625
450 026	**SD**	A	*SW*	NT	63226	64226	68126	63626
450 027	**SD**	A	*SW*	NT	63227	64227	68127	63627
450 028	**SD**	A	*SW*	NT	63228	64228	68128	63628
450 029	**SD**	A	*SW*	NT	63229	64229	68129	63629
450 030	**SD**	A	*SW*	NT	63230	64230	68130	63630
450 031	**SD**	A	*SW*	NT	63231	64231	68131	63631
450 032	**SD**	A	*SW*	NT	63232	64232	68132	63632
450 033	**SD**	A	*SW*	NT	63233	64233	68133	63633
450 034	**SD**	A	*SW*	NT	63234	64234	68134	63634
450 035	**SD**	A	*SW*	NT	63235	64235	68135	63635
450 036	**SD**	A	*SW*	NT	63236	64236	68136	63636
450 037	**SD**	A	*SW*	NT	63237	64237	68137	63637
450 038	**SD**	A	*SW*	NT	63238	64238	68138	63638
450 039	**SD**	A	*SW*	NT	63239	64239	68139	63639
450 040	**SD**	A	*SW*	NT	63240	64240	68140	63640
450 041	**SD**	A	*SW*	NT	63241	64241	68141	63641
450 042	**SD**	A	*SW*	NT	63242	64242	68142	63642
450 071	**SD**	A	*SW*	NT	63271	64271	68171	63671

450 072	**SD**	A	*SW*	NT	63272	64272	68172	63672
450 073	**SD**	A	*SW*	NT	63273	64273	68173	63673
450 074	**SD**	A	*SW*	NT	63274	64274	68174	63674
450 075	**SD**	A	*SW*	NT	63275	64275	68175	63675
450 076	**SD**	A	*SW*	NT	63276	64276	68176	63676
450 077	**SD**	A	*SW*	NT	63277	64277	68177	63677
450 078	**SD**	A	*SW*	NT	63278	64278	68178	63678
450 079	**SD**	A	*SW*	NT	63279	64279	68179	63679
450 080	**SD**	A	*SW*	NT	63280	64280	68180	63680
450 081	**SD**	A	*SW*	NT	63281	64281	68181	63681
450 082	**SD**	A	*SW*	NT	63282	64282	68182	63682
450 083	**SD**	A	*SW*	NT	63283	64283	68183	63683
450 084	**SD**	A	*SW*	NT	63284	64284	68184	63684
450 085	**SD**	A	*SW*	NT	63285	64285	68185	63685
450 086	**SD**	A	*SW*	NT	63286	64286	68186	63686
450 087	**SD**	A	*SW*	NT	63287	64287	68187	63687
450 088	**SD**	A	*SW*	NT	63288	64288	68188	63688
450 089	**SD**	A	*SW*	NT	63289	64289	68189	63689
450 090	**SD**	A	*SW*	NT	63290	64290	68190	63690
450 091	**SD**	A	*SW*	NT	63291	64291	68191	63691
450 092	**SD**	A	*SW*	NT	63292	64292	68192	63692
450 093	**SD**	A	*SW*	NT	63293	64293	68193	63693
450 094	**SD**	A	*SW*	NT	63294	64294	68194	63694
450 095	**SD**	A	*SW*	NT	63295	64295	68195	63695
450 096	**SD**	A	*SW*	NT	63296	64296	68196	63696
450 097	**SD**	A	*SW*	NT	63297	64297	68197	63697
450 098	**SD**	A	*SW*	NT	63298	64298	68198	63698
450 099	**SD**	A	*SW*	NT	63299	64299	68199	63699
450 100	**SD**	A	*SW*	NT	63300	64300	68200	63700
450 101	**SD**	A	*SW*	NT	63701	66851	66801	63751
450 102	**SD**	A	*SW*	NT	63702	66852	66802	63752
450 103	**SD**	A	*SW*	NT	63703	66853	66803	63753
450 104	**SD**	A	*SW*	NT	63704	66854	66804	63754
450 105	**SD**	A	*SW*	NT	63705	66855	66805	63755
450 106	**SD**	A	*SW*	NT	63706	66856	66806	63756
450 107	**SD**	A	*SW*	NT	63707	66857	66807	63757
450 108	**SD**	A	*SW*	NT	63708	66858	66808	63758
450 109	**SD**	A	*SW*	NT	63709	66859	66809	63759
450 110	**SD**	A	*SW*	NT	63710	66860	66810	63760
450 111	**SD**	A	*SW*	NT	63901	66921	66901	63921
450 112	**SD**	A	*SW*	NT	63902	66922	66902	63922
450 113	**SD**	A	*SW*	NT	63903	66923	66903	63923
450 114	**SD**	A	*SW*	NT	63904	66924	66904	63924
450 115	**SD**	A	*SW*	NT	63905	66925	66905	63925
450 116	**SD**	A	*SW*	NT	63906	66926	66906	63926
450 117	**SD**	A	*SW*	NT	63907	66927	66907	63927
450 118	**SD**	A	*SW*	NT	63908	66928	66908	63928
450 119	**SD**	A	*SW*	NT	63909	66929	66909	63929
450 120	**SD**	A	*SW*	NT	63910	66930	66910	63930
450 121	**SD**	A	*SW*	NT	63911	66931	66911	63931
450 122	**SD**	A	*SW*	NT	63912	66932	66912	63932

450 123	**SD**	A	*SW*	NT	63913	66933	66913	63933
450 124	**SD**	A	*SW*	NT	63914	66934	66914	63934
450 125	**SD**	A	*SW*	NT	63915	66935	66915	63935
450 126	**SD**	A	*SW*	NT	63916	66936	66916	63936
450 127	**SD**	A	*SW*	NT	63917	66937	66917	63937

Names (carried on DMSO(B)):

450 015 DESIRO
450 042 TRELOAR COLLEGE
450 114 FAIRBRIDGE investing in the future

Class 450/5. "High density" units. 28 units converted at Bournemouth for Waterloo–Windsor/Weybridge/Hounslow services. First Class removed and modified seating layout with more standing room. Details as Class 450/0 except:

Formation: DMSO–TSO–TSO–DMSO.

DMSO(A). Siemens Uerdingen/Wien (Vienna) 2002–2004. –/64. 48.0 t.
TSO(A). Siemens Uerdingen/Wien (Vienna) 2002–2004. –/56(4) 1T. 35.5 t.
TSO(B). Siemens Uerdingen/Wien (Vienna) 2002–2004. –/56(9) 1TD 2W. 39.8 t.
DMSO(B). Siemens Uerdingen/Wien (Vienna) 2002–2004. –/64. 48.6 t.

450 543	(450 043)	**SD**	A	*SW*	NT	63243	64243	68143	63643
450 544	(450 044)	**SD**	A	*SW*	NT	63244	64244	68144	63644
450 545	(450 045)	**SD**	A	*SW*	NT	63245	64245	68145	63645
450 546	(450 046)	**SD**	A	*SW*	NT	63246	64246	68146	63646
450 547	(450 047)	**SD**	A	*SW*	NT	63247	64247	68147	63647
450 548	(450 048)	**SD**	A	*SW*	NT	63248	64248	68148	63648
450 549	(450 049)	**SD**	A	*SW*	NT	63249	64249	68149	63649
450 550	(450 050)	**SD**	A	*SW*	NT	63250	64250	68150	63650
450 551	(450 051)	**SD**	A	*SW*	NT	63251	64251	68151	63651
450 552	(450 052)	**SD**	A	*SW*	NT	63252	64252	68152	63652
450 553	(450 053)	**SD**	A	*SW*	NT	63253	64253	68153	63653
450 554	(450 054)	**SD**	A	*SW*	NT	63254	64254	68154	63654
450 555	(450 055)	**SD**	A	*SW*	NT	63255	64255	68155	63655
450 556	(450 056)	**SD**	A	*SW*	NT	63256	64256	68156	63656
450 557	(450 057)	**SD**	A	*SW*	NT	63257	64257	68157	63657
450 558	(450 058)	**SD**	A	*SW*	NT	63258	64258	68158	63658
450 559	(450 059)	**SD**	A	*SW*	NT	63259	64259	68159	63659
450 560	(450 060)	**SD**	A	*SW*	NT	63260	64260	68160	63660
450 561	(450 061)	**SD**	A	*SW*	NT	63261	64261	68161	63661
450 562	(450 062)	**SD**	A	*SW*	NT	63262	64262	68162	63662
450 563	(450 063)	**SD**	A	*SW*	NT	63263	64263	68163	63663
450 564	(450 064)	**SD**	A	*SW*	NT	63264	64264	68164	63664
450 565	(450 065)	**SD**	A	*SW*	NT	63265	64265	68165	63665
450 566	(450 066)	**SD**	A	*SW*	NT	63266	64266	68166	63666
450 567	(450 067)	**SD**	A	*SW*	NT	63267	64267	68167	63667
450 568	(450 068)	**SD**	A	*SW*	NT	63268	64268	68168	63668
450 569	(450 069)	**SD**	A	*SW*	NT	63269	64269	68169	63669
450 570	(450 070)	**SD**	A	*SW*	NT	63270	64270	68170	63670

CLASS 455 BR YORK

Inner suburban units.

Formation: DTSO–MSO–TSO–DTSO.
Construction: Steel. Class 455/7 TSO have a steel underframe and an aluminium alloy body & roof.
Traction Motors: Four GEC507-20J of 185 kW, some recovered from Class 405s.
Wheel Arrangement: 2-2 + Bo-Bo + 2-2 + 2-2.
Braking: Disc. **Dimensions:** 19.92/19.83 x 2.82 m.
Bogies: P7 (motor) and T3 (455/8 & 455/9) BX1 (455/7) trailer.
Gangways: Within unit + end doors (sealed on Southern units).
Couplers: Tightlock. **Control System:** 1982-type, camshaft.
Doors: Sliding. **Maximum Speed:** 75 m.p.h.
Heating & Ventilation: Various.
Seating Layout: All units refurbished. SWT units: 2+2 high-back unidirectional/facing seating. Southern units: 3+2 high back mainly facing seating.
Multiple Working: Within class and with Class 456.

Class 455/7. South West Trains units. Second series with TSOs originally in Class 508s. Pressure heating & ventilation.

DTSO. Lot No. 30976 1984–1985. –/50(4) 1W. 30.8 t.
MSO. Lot No. 30975 1984–1985. –/68. 45.7 t.
TSO. Lot No. 30944 1979–1980. –/68. 26.1 t.

5701	**SS**	P	*SW*	WD	77727	62783	71545	77728
5702	**SS**	P	*SW*	WD	77729	62784	71547	77730
5703	**SS**	P	*SW*	WD	77731	62785	71540	77732
5704	**SS**	P	*SW*	WD	77733	62786	71548	77734
5705	**SS**	P	*SW*	WD	77735	62787	71565	77736
5706	**SS**	P	*SW*	WD	77737	62788	71534	77738
5707	**SS**	P	*SW*	WD	77739	62789	71536	77740
5708	**SS**	P	*SW*	WD	77741	62790	71560	77742
5709	**SS**	P	*SW*	WD	77743	62791	71532	77744
5710	**SS**	P	*SW*	WD	77745	62792	71566	77746
5711	**SS**	P	*SW*	WD	77747	62793	71542	77748
5712	**SS**	P	*SW*	WD	77749	62794	71546	77750
5713	**SS**	P	*SW*	WD	77751	62795	71567	77752
5714	**SS**	P	*SW*	WD	77753	62796	71539	77754
5715	**SS**	P	*SW*	WD	77755	62797	71535	77756
5716	**SS**	P	*SW*	WD	77757	62798	71564	77758
5717	**SS**	P	*SW*	WD	77759	62799	71528	77760
5718	**SS**	P	*SW*	WD	77761	62800	71557	77762
5719	**SS**	P	*SW*	WD	77763	62801	71558	77764
5720	**SS**	P	*SW*	WD	77765	62802	71568	77766
5721	**SS**	P	*SW*	WD	77767	62803	71553	77768
5722	**SS**	P	*SW*	WD	77769	62804	71533	77770
5723	**SS**	P	*SW*	WD	77771	62805	71526	77772
5724	**SS**	P	*SW*	WD	77773	62806	71561	77774
5725	**SS**	P	*SW*	WD	77775	62807	71541	77776
5726	**SS**	P	*SW*	WD	77777	62808	71556	77778

5727	**SS**	P	*SW*	WD	77779	62809	71562	77780
5728	**SS**	P	*SW*	WD	77781	62810	71527	77782
5729	**SS**	P	*SW*	WD	77783	62811	71550	77784
5730	**SS**	P	*SW*	WD	77785	62812	71551	77786
5731	**SS**	P	*SW*	WD	77787	62813	71555	77788
5732	**SS**	P	*SW*	WD	77789	62814	71552	77790
5733	**SS**	P	*SW*	WD	77791	62815	71549	77792
5734	**SS**	P	*SW*	WD	77793	62816	71531	77794
5735	**SS**	P	*SW*	WD	77795	62817	71563	77796
5736	**SS**	P	*SW*	WD	77797	62818	71554	77798
5737	**SS**	P	*SW*	WD	77799	62819	71544	77800
5738	**SS**	P	*SW*	WD	77801	62820	71529	77802
5739	**SS**	P	*SW*	WD	77803	62821	71537	77804
5740	**SS**	P	*SW*	WD	77805	62822	71530	77806
5741	**SS**	P	*SW*	WD	77807	62823	71559	77808
5742	**SS**	P	*SW*	WD	77809	62824	71543	77810
5750	**SS**	P	*SW*	WD	77811	62825	71538	77812

Class 455/8. Southern units. First series. Pressure heating & ventilation. Fitted with in-cab air conditioning systems meaning that the end door has been sealed.

DTSO. Lot No. 30972 York 1982–1984. –/74. 33.6 t.
MSO. Lot No. 30973 York 1982–1984. –/84. 37.9 t.
TSO. Lot No. 30974 York 1982–1984. –/75(3) 2W. 34.0 t.

455 801	**SN**	E	*SN*	SU	77627	62709	71657	77580
455 802	**SN**	E	*SN*	SU	77581	62710	71664	77582
455 803	**SN**	E	*SN*	SU	77583	62711	71639	77584
455 804	**SN**	E	*SN*	SU	77585	62712	71640	77586
455 805	**SN**	E	*SN*	SU	77587	62713	71641	77588
455 806	**SN**	E	*SN*	SU	77589	62714	71642	77590
455 807	**SN**	E	*SN*	SU	77591	62715	71643	77592
455 808	**SN**	E	*SN*	SU	77637	62716	71644	77594
455 809	**SN**	E	*SN*	SU	77623	62717	71648	77602
455 810	**SN**	E	*SN*	SU	77597	62718	71646	77598
455 811	**SN**	E	*SN*	SU	77599	62719	71647	77600
455 812	**SN**	E	*SN*	SU	77595	62720	71645	77626
455 813	**SN**	E	*SN*	SU	77603	62721	71649	77604
455 814	**SN**	E	*SN*	SU	77605	62722	71650	77606
455 815	**SN**	E	*SN*	SU	77607	62723	71651	77608
455 816	**SN**	E	*SN*	SU	77609	62724	71652	77633
455 817	**SN**	E	*SN*	SU	77611	62725	71653	77612
455 818	**SN**	E	*SN*	SU	77613	62726	71654	77632
455 819	**SN**	E	*SN*	SU	77615	62727	71637	77616
455 820	**SN**	E	*SN*	SU	77617	62728	71656	77618
455 821	**SN**	E	*SN*	SU	77619	62729	71655	77620
455 822	**SN**	E	*SN*	SU	77621	62730	71658	77622
455 823	**SN**	E	*SN*	SU	77601	62731	71659	77596
455 824	**SN**	E	*SN*	SU	77593	62732	71660	77624
455 825	**SN**	E	*SN*	SU	77579	62733	71661	77628
455 826	**SN**	E	*SN*	SU	77630	62734	71662	77629
455 827	**SN**	E	*SN*	SU	77610	62735	71663	77614

455 828	**SN**	E	*SN*	SU	77631	62736	71638	77634
455 829	**SN**	E	*SN*	SU	77635	62737	71665	77636
455 830	**SN**	E	*SN*	SU	77625	62743	71666	77638
455 831	**SN**	E	*SN*	SU	77639	62739	71667	77640
455 832	**SN**	E	*SN*	SU	77641	62740	71668	77642
455 833	**SN**	E	*SN*	SU	77643	62741	71669	77644
455 834	**SN**	E	*SN*	SU	77645	62742	71670	77646
455 835	**SN**	E	*SN*	SU	77647	62738	71671	77648
455 836	**SN**	E	*SN*	SU	77649	62744	71672	77650
455 837	**SN**	E	*SN*	SU	77651	62745	71673	77652
455 838	**SN**	E	*SN*	SU	77653	62746	71674	77654
455 839	**SN**	E	*SN*	SU	77655	62747	71675	77656
455 840	**SN**	E	*SN*	SU	77657	62748	71676	77658
455 841	**SN**	E	*SN*	SU	77659	62749	71677	77660
455 842	**SN**	E	*SN*	SU	77661	62750	71678	77662
455 843	**SN**	E	*SN*	SU	77663	62751	71679	77664
455 844	**SN**	E	*SN*	SU	77665	62752	71680	77666
455 845	**SN**	E	*SN*	SU	77667	62753	71681	77668
455 846	**SN**	E	*SN*	SU	77669	62754	71682	77670

Class 455/8. South West Trains units. First series. Pressure heating & ventilation.

DTSO. Lot No. 30972 York 1982–1984. –50(4) 1W. 29.5 t.
MSO. Lot No. 30973 York 1982–1984. –/84 –/68. 45.6 t.
TSO. Lot No. 30974 York 1982–1984. –/84 –/68. 27.1 t.

5847	**SS**	P	*SW*	WD	77671	62755	71683	77672
5848	**SS**	P	*SW*	WD	77673	62756	71684	77674
5849	**SS**	P	*SW*	WD	77675	62757	71685	77676
5850	**SS**	P	*SW*	WD	77677	62758	71686	77678
5851	**SS**	P	*SW*	WD	77679	62759	71687	77680
5852	**SS**	P	*SW*	WD	77681	62760	71688	77682
5853	**SS**	P	*SW*	WD	77683	62761	71689	77684
5854	**SS**	P	*SW*	WD	77685	62762	71690	77686
5855	**SS**	P	*SW*	WD	77687	62763	71691	77688
5856	**SS**	P	*SW*	WD	77689	62764	71692	77690
5857	**SS**	P	*SW*	WD	77691	62765	71693	77692
5858	**SS**	P	*SW*	WD	77693	62766	71694	77694
5859	**SS**	P	*SW*	WD	77695	62767	71695	77696
5860	**SS**	P	*SW*	WD	77697	62768	71696	77698
5861	**SS**	P	*SW*	WD	77699	62769	71697	77700
5862	**SS**	P	*SW*	WD	77701	62770	71698	77702
5863	**SS**	P	*SW*	WD	77703	62771	71699	77704
5864	**SS**	P	*SW*	WD	77705	62772	71700	77706
5865	**SS**	P	*SW*	WD	77707	62773	71701	77708
5866	**SS**	P	*SW*	WD	77709	62774	71702	77710
5867	**SS**	P	*SW*	WD	77711	62775	71703	77712
5868	**SS**	P	*SW*	WD	77713	62776	71704	77714
5869	**SS**	P	*SW*	WD	77715	62777	71705	77716
5870	**SS**	P	*SW*	WD	77717	62778	71706	77718
5871	**SS**	P	*SW*	WD	77719	62779	71707	77720
5872	**SS**	P	*SW*	WD	77721	62780	71708	77722

5873	**SS**	P	*SW*	WD	77723	62781	71709	77724
5874	**SS**	P	*SW*	WD	77725	62782	71710	77726

Class 455/9. South West Trains units. Third series. Convection heating. **Dimensions**: 19.96/20.18 x 2.82 m.

DTSO. Lot No. 30991 York 1985. –/50(4) 1W. 30.7 t.
MSO. Lot No. 30992 York 1985. –/68. 46.3 t.
TSO. Lot No. 30993 York 1985. –/68. 28.3 t.
TSO†. Lot No. 30932 Derby 1981. –/68. 26.5 t.

Note: † Prototype vehicle 67400 converted from a Class 210 DEMU.

5901		**SS**	P	*SW*	WD	77813	62826	71714	77814
5902		**SS**	P	*SW*	WD	77815	62827	71715	77816
5903		**SS**	P	*SW*	WD	77817	62828	71716	77818
5904		**SS**	P	*SW*	WD	77819	62829	71717	77820
5905		**SS**	P	*SW*	WD	77821	62830	71725	77822
5906		**SS**	P	*SW*	WD	77823	62831	71719	77824
5907		**SS**	P	*SW*	WD	77825	62832	71720	77826
5908		**SS**	P	*SW*	WD	77827	62833	71721	77828
5909		**SS**	P	*SW*	WD	77829	62834	71722	77830
5910		**SS**	P	*SW*	WD	77831	62835	71723	77832
5911		**SS**	P	*SW*	WD	77833	62836	71724	77834
5912	†	**SS**	P	*SW*	WD	77835	62837	67400	77836
5913		**SS**	P	*SW*	WD	77837	62838	71726	77838
5914		**SS**	P	*SW*	WD	77839	62839	71727	77840
5915		**SS**	P	*SW*	WD	77841	62840	71728	77842
5916		**SS**	P	*SW*	WD	77843	62841	71729	77844
5917		**SS**	P	*SW*	WD	77845	62842	71730	77846
5918		**SS**	P	*SW*	WD	77847	62843	71732	77848
5919		**SS**	P	*SW*	WD	77849	62844	71718	77850
5920		**SS**	P	*SW*	WD	77851	62845	71733	77852

CLASS 456 BREL YORK

Inner suburban units.

Formation: DMSO–DTSO.
Construction: Steel underframe, aluminium alloy body & roof.
Traction Motors: Two GEC507-20J of 185 kW, some recovered from Class 405s.
Wheel Arrangement: 2-Bo + 2-2.
Braking: Disc.
Bogies: P7 (motor) and T3 (trailer).
Gangways: Within unit.
Doors: Sliding.
Dimensions: 20.61 x 2.82 m.
Couplers: Tightlock.
Control System: GTO Chopper.
Maximum Speed: 75 m.p.h.
Seating Layout: 3+2 facing.
Heating & Ventilation: Convection heating.
Multiple Working: Within class and with Class 455.

DMSO. Lot No. 31073 1990–1991. –/79. 41.1 t.
DTSO. Lot No. 31074 1990–1991. –/73. 31.4 t.

Advertising livery: 456 006 TfL/City of London (blue & green with various images).

456 001	**SN**	P	*SN*	SU	64735	78250	
456 002	**SN**	P	*SN*	SU	64736	78251	
456 003	**SN**	P	*SN*	SU	64737	78252	
456 004	**SN**	P	*SN*	SU	64738	78253	
456 005	**SN**	P	*SN*	SU	64739	78254	
456 006	**AL**	P	*SN*	SU	64740	78255	
456 007	**SN**	P	*SN*	SU	64741	78256	
456 008	**SN**	P	*SN*	SU	64742	78257	
456 009	**SN**	P	*SN*	SU	64743	78258	
456 010	**SN**	P	*SN*	SU	64744	78259	
456 011	**SN**	P	*SN*	SU	64745	78260	
456 012	**SN**	P	*SN*	SU	64746	78261	
456 013	**SN**	P	*SN*	SU	64747	78262	
456 014	**SN**	P	*SN*	SU	64748	78263	
456 015	**SN**	P	*SN*	SU	64749	78264	
456 016	**SN**	P	*SN*	SU	64750	78265	
456 017	**SN**	P	*SN*	SU	64751	78266	
456 018	**SN**	P	*SN*	SU	64752	78267	
456 019	**SN**	P	*SN*	SU	64753	78268	
456 020	**SN**	P	*SN*	SU	64754	78269	
456 021	**SN**	P	*SN*	SU	64755	78270	
456 022	**SN**	P	*SN*	SU	64756	78271	
456 023	**SN**	P	*SN*	SU	64757	78272	
456 024	**SN**	P	*SN*	SU	64758	78273	Sir Cosmo Bonsor

CLASS 458 JUNIPER ALSTOM BIRMINGHAM

Outer suburban units.

Formation: DMCO–TSO–MSO–DMCO.
Construction: Steel.
Traction Motors: Two Alstom ONIX 800 asynchronous of 270 kW.
Wheel Arrangement: 2-Bo + 2-2 + Bo-2 + Bo-2.
Braking: Disc & regenerative.
Dimensions: 21.16/19.94 x 2.80 m.
Bogies: ACR.
Couplers: Scharfenberg AAR.
Gangways: Throughout (not in use).
Control System: IGBT Inverter.
Doors: Sliding plug.
Maximum Speed: 100 m.p.h.
Heating & Ventilation: Air conditioning.
Multiple Working: Within class.
Seating Layout: 1: 2+2 facing, 2: 3+2 facing/unidirectional.

DMCO(A). Alstom 1998–2000. 12/63. 46.4 t.
TSO. Alstom 1998–2000. –/54(6) 1TD 2W. 34.6 t.
MSO. Alstom 1998–2000. –/75 1T. 42.1 t.
DMCO(B). Alstom 1998–2000. 12/63. 46.4 t.

8001	**ST**	P	*SW*	WD	67601	74001	74101	67701
8002	**ST**	P	*SW*	WD	67602	74002	74102	67702
8003	**ST**	P	*SW*	WD	67603	74003	74103	67703
8004	**ST**	P	*SW*	WD	67604	74004	74104	67704
8005	**ST**	P	*SW*	WD	67605	74005	74105	67705
8006	**ST**	P	*SW*	WD	67606	74006	74106	67706
8007	**ST**	P	*SW*	WD	67607	74007	74107	67707

8008	**ST**	P	*SW*	WD	67608	74008	74108	67708
8009	**ST**	P	*SW*	WD	67609	74009	74109	67709
8010	**ST**	P	*SW*	WD	67610	74010	74110	67710
8011	**ST**	P	*SW*	WD	67611	74011	74111	67711
8012	**ST**	P	*SW*	WD	67612	74012	74112	67712
8013	**ST**	P	*SW*	WD	67613	74013	74113	67713
8014	**ST**	P	*SW*	WD	67614	74014	74114	67714
8015	**ST**	P	*SW*	WD	67615	74015	74115	67715
8016	**ST**	P	*SW*	WD	67616	74016	74116	67716
8017	**ST**	P	*SW*	WD	67617	74017	74117	67717
8018	**ST**	P	*SW*	WD	67618	74018	74118	67718
8019	**ST**	P	*SW*	WD	67619	74019	74119	67719
8020	**ST**	P	*SW*	WD	67620	74020	74120	67720
8021	**ST**	P	*SW*	WD	67621	74021	74121	67721
8022	**ST**	P	*SW*	WD	67622	74022	74122	67722
8023	**ST**	P	*SW*	WD	67623	74023	74123	67723
8024	**ST**	P	*SW*	WD	67624	74024	74124	67724
8025	**ST**	P	*SW*	WD	67625	74025	74125	67725
8026	**ST**	P	*SW*	WD	67626	74026	74126	67726
8027	**ST**	P	*SW*	WD	67627	74027	74127	67727
8028	**ST**	P	*SW*	WD	67628	74028	74128	67728
8029	**ST**	P	*SW*	WD	67629	74029	74129	67729
8030	**ST**	P	*SW*	WD	67630	74030	74130	67730

CLASS 460 GEC-ALSTHOM JUNIPER

Only the last two digits of the unit number are carried on the front ends of these units. Used on London Victoria–Gatwick Airport express services.

Formation: DMLFO–TFO–TCO–MSO–MSO–TSO–MSO–DMSO.
Construction: Steel.
Traction Motors: Two Alstom ONIX 800 asynchronous of 270 kW.
Wheel Arrangement: 2-Bo + 2-2 + 2-2 +Bo-2 + 2-Bo + 2-2 + Bo-2 + Bo-2.
Braking: Disc & regenerative. **Dimensions**: 21.01/19.94 x 2.80 m.
Bogies: ACR.
Couplers: Scharfenberg 330 at outer ends and between cars 4 and 5.
Gangways: Within unit. **Control System**: IGBT Inverter.
Doors: Sliding plug. **Maximum Speed**: 100 m.p.h.
Heating & Ventilation: Air conditioning.
Seating Layout: 1: 2+1 facing, 2: 2+2 facing/unidirectional.
Multiple Working: Within class.

DMLFO. Alstom 1998–1999. 10/– 42.7 t.
TFO. Alstom 1998–1999. 25/– 1TD 1W. 34.5 t.
TCO. Alstom 1998–1999. 8/38 1T. 35.6 t.
MSO(A). Alstom 1998–1999. –/58. 42.8 t.
MSO(B). Alstom 1998–1999. –/58. 42.5 t.
TSO. Alstom 1998–1999. –/33 1TD 1T 1W. 35.2 t.
MSO(C). Alstom 1998–1999. –/58. 40.5 t.
DMSO. Alstom 1998–1999. –/54. 45.4 t.

Advertising liveries:

460 002	Emirates Airlines (Australia)	460 004	Emirates Airlines (General)
460 003	Emirates Airlines (China)	460 006	Emirates Airlines (Africa)

460 001	**GV**	P	*SN*	SL	67901	74401	74411	74421
					74431	74441	74451	67911
460 002	**AL**	P	*SN*	SL	67902	74402	74412	74422
					74432	74442	74452	67912
460 003	**AL**	P	*SN*	SL	67903	74403	74413	74423
					74433	74443	74453	67913
460 004	**AL**	P	*SN*	SL	67904	74404	74414	74424
					74434	74444	74454	67914
460 005	**GV**	P	*SN*	SL	67905	74405	74415	74425
					74435	74445	74455	67915
460 006	**AL**	P	*SN*	SL	67906	74406	74416	74426
					74436	74446	74456	67916
460 007	**GV**	P	*SN*	SL	67907	74407	74417	74427
					74437	74447	74457	67917
460 008	**GV**	P	*SN*	SL	67908	74408	74418	74428
					74438	74448	74458	67918

CLASS 465 NETWORKER

Inner/outer suburban units.

Formation: DMSO–TSO–TSO–DMSO.
Construction: Welded aluminium alloy.
Traction Motors: Hitachi asynchronous of 280 kW (Classes 465/0 and 465/1) or GEC-Alsthom G352BY (Classes 465/2 and 465/9).
Wheel Arrangement: Bo-Bo + 2-2 + 2-2 + Bo-Bo.
Braking: Disc & rheostatic and regenerative (Classes 465/0 and 465/1 only).
Bogies: BREL P3/T3 (465/0 and 465/1), SRP BP62/BT52 (465/2 and 465/9).
Dimensions: 20.89/20.06 x 2.81 m.
Control System: IGBT Inverter (465/0 and 465/1) or 1992-type GTO Inverter.
Gangways: Within unit. **Couplers**: Tightlock.
Doors: Sliding plug. **Maximum Speed**: 75 m.p.h.
Seating Layout: 3+2 facing/unidirectional.
Multiple Working: Within class and with Class 466.

64759–64808. DMSO(A). Lot No. 31100 BREL York 1991–1993. –/86. 39.2 t.
64809–64858. DMSO(B). Lot No. 31100 BREL York 1991–1993. –/86. 39.2 t.
65734–65749. DMSO(A). Lot No. 31103 Metro-Cammell 1991–1993. –/86. 39.2 t.
65784–65799. DMSO(B). Lot No. 31103 Metro-Cammell 1991–1993. –/86. 39.2 t.
65800–65846. DMSO(A). Lot No. 31130 ABB York 1993–1994. –/86. 39.2 t.
65847–65893. DMSO(B). Lot No. 31130 ABB York 1993–1994. –/86. 39.2 t.
72028–72126 (even nos.) TSO. Lot No. 31102 BREL York 1991–1993. –/90. 27.2 t.
72029–72127 (odd nos.) TSO. Lot No. 31101 BREL York 1991–1993. –/86 1T. 28.0 t.
72787–72817 (odd nos.) TSO. Lot No. 31104 Metro-Cammell 1991–1992. –/86 1T. 28.0 t.
72788–72818 (even nos.) TSO. Lot No. 31105 Metro-Cammell 1991–1992. –/90. 27.2 t.
72900–72992 (even nos.) TSO. Lot No. 31102 ABB York 1993–1994. –/90. 27.2 t.
72901–72993 (odd nos.) TSO. Lot No. 31101 ABB York 1993–1994. –/86 1T. 28.0 t.

Class 465/0. Built by BREL/ABB.

465 001	**CN**	E	*SE*	SG	64759	72028	72029	64809
465 002	**CN**	E	*SE*	SG	64760	72030	72031	64810
465 003	**CN**	E	*SE*	SG	64761	72032	72033	64811
465 004	**SE**	E	*SE*	SG	64762	72034	72035	64812
465 005	**CN**	E	*SE*	SG	64763	72036	72037	64813
465 006	**CN**	E	*SE*	SG	64764	72038	72039	64814
465 007	**CN**	E	*SE*	SG	64765	72040	72041	64815
465 008	**SE**	E	*SE*	SG	64766	72042	72043	64816
465 009	**SE**	E	*SE*	SG	64767	72044	72045	64817
465 010	**SE**	E	*SE*	SG	64768	72046	72047	64818
465 011	**CN**	E	*SE*	SG	64769	72048	72049	64819
465 012	**SE**	E	*SE*	SG	64770	72050	72051	64820
465 013	**CN**	E	*SE*	SG	64771	72052	72053	64821
465 014	**CN**	E	*SE*	SG	64772	72054	72055	64822
465 015	**CN**	E	*SE*	SG	64773	72056	72057	64823
465 016	**SE**	E	*SE*	SG	64774	72058	72059	64824
465 017	**SE**	E	*SE*	SG	64775	72060	72061	64825
465 018	**CN**	E	*SE*	SG	64776	72062	72063	64826
465 019	**SE**	E	*SE*	SG	64777	72064	72065	64827
465 020	**SE**	E	*SE*	SG	64778	72066	72067	64828
465 021	**CN**	E	*SE*	SG	64779	72068	72069	64829
465 022	**CN**	E	*SE*	SG	64780	72070	72071	64830
465 023	**CN**	E	*SE*	SG	64781	72072	72073	64831
465 024	**SE**	E	*SE*	SG	64782	72074	72075	64832
465 025	**SE**	E	*SE*	SG	64783	72076	72077	64833
465 026	**SE**	E	*SE*	SG	64784	72078	72079	64834
465 027	**CN**	E	*SE*	SG	64785	72080	72081	64835
465 028	**CN**	E	*SE*	SG	64786	72082	72083	64836
465 029	**CN**	E	*SE*	SG	64787	72084	72085	64837
465 030	**CN**	E	*SE*	SG	64788	72086	72087	64838
465 031	**SE**	E	*SE*	SG	64789	72088	72089	64839
465 032	**CN**	E	*SE*	SG	64790	72090	72091	64840
465 033	**CN**	E	*SE*	SG	64791	72092	72093	64841
465 034	**SE**	E	*SE*	SG	64792	72094	72095	64842
465 035	**CN**	E	*SE*	SG	64793	72096	72097	64843
465 036	**SE**	E	*SE*	SG	64794	72098	72099	64844
465 037	**SE**	E	*SE*	SG	64795	72100	72101	64845
465 038	**CN**	E	*SE*	SG	64796	72102	72103	64846
465 039	**CN**	E	*SE*	SG	64797	72104	72105	64847
465 040	**CN**	E	*SE*	SG	64798	72106	72107	64848
465 041	**CN**	E	*SE*	SG	64799	72108	72109	64849
465 042	**CN**	E	*SE*	SG	64800	72110	72111	64850
465 043	**CN**	E	*SE*	SG	64801	72112	72113	64851
465 044	**CN**	E	*SE*	SG	64802	72114	72115	64852
465 045	**CN**	E	*SE*	SG	64803	72116	72117	64853
465 046	**CN**	E	*SE*	SG	64804	72118	72119	64854
465 047	**CN**	E	*SE*	SG	64805	72120	72121	64855
465 048	**CN**	E	*SE*	SG	64806	72122	72123	64856
465 049	**SE**	E	*SE*	SG	64807	72124	72125	64857

465 050	**CN**	E	*SE*	SG	64808	72126	72127	64858

Class 465/1. Built by BREL/ABB. Similar to Class 465/0 but with detail differences.

465 151	**CN**	E	*SE*	SG	65800	72900	72901	65847
465 152	**CN**	E	*SE*	SG	65801	72902	72903	65848
465 153	**CN**	E	*SE*	SG	65802	72904	72905	65849
465 154	**SE**	E	*SE*	SG	65803	72906	72907	65850
465 155	**CN**	E	*SE*	SG	65804	72908	72909	65851
465 156	**CN**	E	*SE*	SG	65805	72910	72911	65852
465 157	**CN**	E	*SE*	SG	65806	72912	72913	65853
465 158	**CN**	E	*SE*	SG	65807	72914	72915	65854
465 159	**CN**	E	*SE*	SG	65808	72916	72917	65855
465 160	**CN**	E	*SE*	SG	65809	72918	72919	65856
465 161	**CN**	E	*SE*	SG	65810	72920	72921	65857
465 162	**CN**	E	*SE*	SG	65811	72922	72923	65858
465 163	**CN**	E	*SE*	SG	65812	72924	72925	65859
465 164	**CN**	E	*SE*	SG	65813	72926	72927	65860
465 165	**CN**	E	*SE*	SG	65814	72928	72929	65861
465 166	**CN**	E	*SE*	SG	65815	72930	72931	65862
465 167	**CN**	E	*SE*	SG	65816	72932	72933	65863
465 168	**CN**	E	*SE*	SG	65817	72934	72935	65864
465 169	**CN**	E	*SE*	SG	65818	72936	72937	65865
465 170	**CN**	E	*SE*	SG	65819	72938	72939	65866
465 171	**CN**	E	*SE*	SG	65820	72940	72941	65867
465 172	**CN**	E	*SE*	SG	65821	72942	72943	65868
465 173	**CN**	E	*SE*	SG	65822	72944	72945	65869
465 174	**CN**	E	*SE*	SG	65823	72946	72947	65870
465 175	**CN**	E	*SE*	SG	65824	72948	72949	65871
465 176	**CN**	E	*SE*	SG	65825	72950	72951	65872
465 177	**SE**	E	*SE*	SG	65826	72952	72953	65873
465 178	**CN**	E	*SE*	SG	65827	72954	72955	65874
465 179	**CN**	E	*SE*	SG	65828	72956	72957	65875
465 180	**CN**	E	*SE*	SG	65829	72958	72959	65876
465 181	**CN**	E	*SE*	SG	65830	72960	72961	65877
465 182	**CN**	E	*SE*	SG	65831	72962	72963	65878
465 183	**CN**	E	*SE*	SG	65832	72964	72965	65879
465 184	**CN**	E	*SE*	SG	65833	72966	72967	65880
465 185	**CN**	E	*SE*	SG	65834	72968	72969	65881
465 186	**CN**	E	*SE*	SG	65835	72970	72971	65882
465 187	**CN**	E	*SE*	SG	65836	72972	72973	65883
465 188	**CN**	E	*SE*	SG	65837	72974	72975	65884
465 189	**CN**	E	*SE*	SG	65838	72976	72977	65885
465 190	**CN**	E	*SE*	SG	65839	72978	72979	65886
465 191	**CN**	E	*SE*	SG	65840	72980	72981	65887
465 192	**CN**	E	*SE*	SG	65841	72982	72983	65888
465 193	**CN**	E	*SE*	SG	65842	72984	72985	65889
465 194	**CN**	E	*SE*	SG	65843	72986	72987	65890
465 195	**CN**	E	*SE*	SG	65844	72988	72989	65891
465 196	**CN**	E	*SE*	SG	65845	72990	72991	65892
465 197	**CN**	E	*SE*	SG	65846	72992	72993	65893

Class 465/2. Built by Metro-Cammell.
Dimensions: 20.80/20.15 x 2.81 m.

465 235	**SE**	A	*SE*	SG	65734	72787	72788	65784
465 236	**CN**	A	*SE*	SG	65735	72789	72790	65785
465 237	**CN**	A	*SE*	SG	65736	72791	72792	65786
465 238	**CN**	A	*SE*	SG	65737	72793	72794	65787
465 239	**CN**	A	*SE*	SG	65738	72795	72796	65788
465 240	**SE**	A	*SE*	SG	65739	72797	72798	65789
465 241	**SE**	A	*SE*	SG	65740	72799	72800	65790
465 242	**CN**	A	*SE*	SG	65741	72801	72802	65791
465 243	**SE**	A	*SE*	SG	65742	72803	72804	65792
465 244	**CN**	A	*SE*	SG	65743	72805	72806	65793
465 245	**CN**	A	*SE*	SG	65744	72807	72808	65794
465 246	**CN**	A	*SE*	SG	65745	72809	72810	65795
465 247	**CN**	A	*SE*	SG	65746	72811	72812	65796
465 248	**CN**	A	*SE*	SG	65747	72813	72814	65797
465 249	**CN**	A	*SE*	SG	65748	72815	72816	65798
465 250	**CN**	A	*SE*	SG	65749	72817	72818	65799

Class 465/9. Built by Metro-Cammell. Refurbished 2005 for longer distance services, with the addition of First Class seats. Details as Class 465/0 unless stated.
Formation: DMCO–TSO(A)–TSO(B)–DMCO.
Seating Layout: 1: 2+2 facing/unidirectional, 2: 3+2 facing/unidirectional.

65700–65733. DMCO(A). Lot No. 31103 Metro-Cammell 1991–1993. 12/68. 39.2 t.
72719–72785 (odd nos.) TSO(A). Lot No. 31104 Metro-Cammell 1991–1992. –/76 1T 2W. 30.3 t.
72720–72786 (even nos.) TSO(B). Lot No. 31105 Metro-Cammell 1991–1992. –/90. 29.5 t.
65750–65783. DMCO(B). Lot No. 31103 Metro-Cammell 1991–1993. 12/68. 39.2 t.

465 901	(465 201)	**CN**	A	*SE*	SG	65700	72719	72720	65750
465 902	(465 202)	**CN**	A	*SE*	SG	65701	72721	72722	65751
465 903	(465 203)	**CN**	A	*SE*	SG	65702	72723	72724	65752
465 904	(465 204)	**CN**	A	*SE*	SG	65703	72725	72726	65753
465 905	(465 205)	**CN**	A	*SE*	SG	65704	72727	72728	65754
465 906	(465 206)	**CN**	A	*SE*	SG	65705	72729	72730	65755
465 907	(465 207)	**CN**	A	*SE*	SG	65706	72731	72732	65756
465 908	(465 208)	**CN**	A	*SE*	SG	65707	72733	72734	65757
465 909	(465 209)	**CN**	A	*SE*	SG	65708	72735	72736	65758
465 910	(465 210)	**CN**	A	*SE*	SG	65709	72737	72738	65759
465 911	(465 211)	**CN**	A	*SE*	SG	65710	72739	72740	65760
465 912	(465 212)	**CN**	A	*SE*	SG	65711	72741	72742	65761
465 913	(465 213)	**SE**	A	*SE*	SG	65712	72743	72744	65762
465 914	(465 214)	**CN**	A	*SE*	SG	65713	72745	72746	65763
465 915	(465 215)	**CN**	A	*SE*	SG	65714	72747	72748	65764
465 916	(465 216)	**CN**	A	*SE*	SG	65715	72749	72750	65765
465 917	(465 217)	**CN**	A	*SE*	SG	65716	72751	72752	65766
465 918	(465 218)	**CN**	A	*SE*	SG	65717	72753	72754	65767
465 919	(465 219)	**CN**	A	*SE*	SG	65718	72755	72756	65768
465 920	(465 220)	**CN**	A	*SE*	SG	65719	72757	72758	65769
465 921	(465 221)	**CN**	A	*SE*	SG	65720	72759	72760	65770

465 922	(465 222)	**CN**	A	*SE*	SG	65721	72761	72762	65771
465 923	(465 223)	**CN**	A	*SE*	SG	65722	72763	72764	65772
465 924	(465 224)	**SE**	A	*SE*	SG	65723	72765	72766	65773
465 925	(465 225)	**CN**	A	*SE*	SG	65724	72767	72768	65774
465 926	(465 226)	**CN**	A	*SE*	SG	65725	72769	72770	65775
465 927	(465 227)	**CN**	A	*SE*	SG	65726	72771	72772	65776
465 928	(465 228)	**CN**	A	*SE*	SG	65727	72773	72774	65777
465 929	(465 229)	**CN**	A	*SE*	SG	65728	72775	72776	65778
465 930	(465 230)	**CN**	A	*SE*	SG	65729	72777	72778	65779
465 931	(465 231)	**CN**	A	*SE*	SG	65730	72779	72780	65780
465 932	(465 232)	**CN**	A	*SE*	SG	65731	72781	72782	65781
465 933	(465 233)	**CN**	A	*SE*	SG	65732	72783	72784	65782
465 934	(465 234)	**CN**	A	*SE*	SG	65733	72785	72786	65783

Name: 465 903 Remembrance

CLASS 466 NETWORKER GEC-ALSTHOM

Inner/outer suburban units.

Formation: DMSO–DTSO.
Construction: Welded aluminium alloy.
Traction Motors: Two GEC-Alsthom G352AY asynchronous of 280 kW.
Wheel Arrangement: Bo-Bo + 2-2.
Couplers: Tightlock.
Braking: Disc & rheostatic.
Control System: 1992-type GTO Inverter.
Dimensions: 20.80 x 2.80 m.
Maximum Speed: 75 m.p.h.
Bogies: BREL P3/T3.
Doors: Sliding plug.
Gangways: Within unit.
Seating Layout: 3+2 facing/unidirectional.
Multiple Working: Within class and with Class 465.

DMSO. Lot No. 31128 Birmingham 1993–1994. –/86. 40.6 t.
DTSO. Lot No. 31129 Birmingham 1993–1994. –/82 1T. 31.4 t.

466 001	**CN**	A	*SE*	SG	64860	78312
466 002	**CN**	A	*SE*	SG	64861	78313
466 003	**CN**	A	*SE*	SG	64862	78314
466 004	**CN**	A	*SE*	SG	64863	78315
466 005	**CN**	A	*SE*	SG	64864	78316
466 006	**CN**	A	*SE*	SG	64865	78317
466 007	**CN**	A	*SE*	SG	64866	78318
466 008	**CN**	A	*SE*	SG	64867	78319
466 009	**CN**	A	*SE*	SG	64868	78320
466 010	**CN**	A	*SE*	SG	64869	78321
466 011	**CN**	A	*SE*	SG	64870	78322
466 012	**CN**	A	*SE*	SG	64871	78323
466 013	**CN**	A	*SE*	SG	64872	78324
466 014	**CN**	A	*SE*	SG	64873	78325
466 015	**CN**	A	*SE*	SG	64874	78326
466 016	**CN**	A	*SE*	SG	64875	78327
466 017	**CN**	A	*SE*	SG	64876	78328
466 018	**CN**	A	*SE*	SG	64877	78329
466 019	**CN**	A	*SE*	SG	64878	78330
466 020	**CN**	A	*SE*	SG	64879	78331

466 021	**CN**	A	*SE*	SG	64880	78332
466 022	**CN**	A	*SE*	SG	64881	78333
466 023	**CN**	A	*SE*	SG	64882	78334
466 024	**CN**	A	*SE*	SG	64883	78335
466 025	**CN**	A	*SE*	SG	64884	78336
466 026	**CN**	A	*SE*	SG	64885	78337
466 027	**CN**	A	*SE*	SG	64886	78338
466 028	**CN**	A	*SE*	SG	64887	78339
466 029	**CN**	A	*SE*	SG	64888	78340
466 030	**CN**	A	*SE*	SG	64889	78341
466 031	**CN**	A	*SE*	SG	64890	78342
466 032	**CN**	A	*SE*	SG	64891	78343
466 033	**CN**	A	*SE*	SG	64892	78344
466 034	**CN**	A	*SE*	SG	64893	78345
466 035	**CN**	A	*SE*	SG	64894	78346
466 036	**CN**	A	*SE*	SG	64895	78347
466 037	**CN**	A	*SE*	SG	64896	78348
466 038	**CN**	A	*SE*	SG	64897	78349
466 039	**CN**	A	*SE*	SG	64898	78350
466 040	**CN**	A	*SE*	SG	64899	78351
466 041	**CN**	A	*SE*	SG	64900	78352
466 042	**CN**	A	*SE*	SG	64901	78353
466 043	**CN**	A	*SE*	SG	64902	78354

CLASS 483 METRO-CAMMELL

Built 1938 onwards for LTE. Converted 1989–1990 for the Isle of Wight Line.

Formation: DMSO–DMSO.
System: 660 V DC third rail.
Construction: Steel.
Traction Motors: Two Crompton Parkinson/GEC/BTH LT100 of 125 kW.
Braking: Tread. **Dimensions**: 16.15 x 2.69 m.
Bogies: LT design. **Couplers**: Wedglock.
Gangways: None. End doors.
Control System: Pneumatic Camshaft Motor (PCM).
Doors: Sliding. **Maximum Speed**: 45 m.p.h.
Seating Layout: Longitudinal or 2+2 facing/unidirectional.
Multiple Working: Within class.

Notes: The last three numbers of the unit number only are carried.

Former London Underground numbers are shown in parentheses.

DMSO (A). Lot No. 31071. –/40. 27.4 t.
DMSO (B). Lot No. 31072. –/42. 27.4 t.

483 002	**LT**	SW	*SW*	RY	122	(10221)	225	(11142)
483 004	**LT**	SW	*SW*	RY	124	(10205)	224	(11205)
483 006	**LT**	SW	*SW*	RY	126	(10297)	226	(11297)
483 007	**LT**	SW	*SW*	RY	127	(10291)	227	(11291)
483 008	**LT**	SW	*SW*	RY	128	(10255)	228	(11255)
483 009	**LT**	SW	*SW*	RY	129	(10289)	229	(11229)

CLASS 507 BREL YORK

Suburban units.

Formation: BDMSO–TSO–DMSO.
Construction: Steel underframe, aluminium alloy body and roof.
Traction Motors: Four GEC G310AZ of 82.125 kW.
Wheel Arrangement: Bo-Bo + 2-2 + Bo-Bo.
Braking: Disc & rheostatic.
Dimensions: 20.18 x 2.82 m.
Bogies: BX1.
Couplers: Tightlock.
Gangways: Within unit + end doors.
Control System: Camshaft.
Doors: Sliding.
Maximum Speed: 75 m.p.h.
Seating Layout: All refurbished with 2+2 high-back facing seating.
Multiple Working: Within class and with Class 508.

BDMSO. Lot No. 30906 1978–1980. –/56(3) 1W. 37.0 t.
TSO. Lot No. 30907 1978–1980. –/74. 25.5 t.
DMSO. Lot No. 30908 1978–1980. –/56(3) 1W. 35.5 t.

507 001	**ME**	A	*ME*	BD	64367	71342	64405	
507 002	**ME**	A	*ME*	BD	64368	71343	64406	
507 003	**ME**	A	*ME*	BD	64369	71344	64407	
507 004	**ME**	A	*ME*	BD	64388	71345	64408	Bob Paisley
507 005	**ME**	A	*ME*	BD	64371	71346	64409	
507 006	**ME**	A	*ME*	BD	64372	71347	64410	
507 007	**ME**	A	*ME*	BD	64373	71348	64411	
507 008	**ME**	A	*ME*	BD	64374	71349	64412	
507 009	**ME**	A	*ME*	BD	64375	71350	64413	Dixie Dean
507 010	**ME**	A	*ME*	BD	64376	71351	64414	
507 011	**ME**	A	*ME*	BD	64377	71352	64415	
507 012	**ME**	A	*ME*	BD	64378	71353	64416	
507 013	**ME**	A	*ME*	BD	64379	71354	64417	
507 014	**ME**	A	*ME*	BD	64380	71355	64418	
507 015	**ME**	A	*ME*	BD	64381	71356	64419	
507 016	**ME**	A	*ME*	BD	64382	71357	64420	
507 017	**ME**	A	*ME*	BD	64383	71358	64421	
507 018	**ME**	A	*ME*	BD	64384	71359	64422	
507 019	**ME**	A	*ME*	BD	64385	71360	64423	
507 020	**ME**	A	*ME*	BD	64386	71361	64424	John Peel
507 021	**ME**	A	*ME*	BD	64387	71362	64425	Red Rum
507 023	**ME**	A	*ME*	BD	64389	71364	64427	
507 024	**ME**	A	*ME*	BD	64390	71365	64428	
507 025	**ME**	A	*ME*	BD	64391	71366	64429	
507 026	**ME**	A	*ME*	BD	64392	71367	64430	
507 027	**ME**	A	*ME*	BD	64393	71368	64431	
507 028	**ME**	A	*ME*	BD	64394	71369	64432	
507 029	**ME**	A	*ME*	BD	64395	71370	64433	
507 030	**ME**	A	*ME*	BD	64396	71371	64434	
507 031	**ME**	A	*ME*	BD	64397	71372	64435	
507 032	**ME**	A	*ME*	BD	64398	71373	64436	
507 033	**ME**	A	*ME*	BD	64399	71374	64437	

CLASS 508 BREL YORK

Suburban units.

Formation: DMSO–TSO–BDMSO.
Construction: Steel underframe, aluminium alloy body and roof.
Traction Motors: Four GEC G310AZ of 82.125 kW.
Wheel Arrangement: Bo-Bo + 2-2 + Bo-Bo.
Braking: Disc & rheostatic.
Dimensions: 20.18 x 2.82 m.
Bogies: BX1.
Couplers: Tightlock.
Gangways: Within unit + end doors.
Control System: Camshaft.
Doors: Sliding.
Maximum Speed: 75 m.p.h.
Seating Layout: All Merseyrail units refurbished with 2+2 high-back facing seating. 508/2 and 508/3 units have 3+2 low-back facing seating.
Multiple Working: Within class and with Class 507.

DMSO. Lot No. 30979 1979–1980. –/56(3) 1W. 36.0 t.
TSO. Lot No. 30980 1979–1980. –/74. 26.5 t.
BDMSO. Lot No. 30981 1979–1980. –/56(3) 1W. 36.5 t.

Class 508/1. Merseyrail units.

508 103	**ME**	A	*ME*	BD	64651	71485	64694
508 104	**ME**	A	*ME*	BD	64652	71486	64695
508 108	**ME**	A	*ME*	BD	64656	71490	64699
508 110	**ME**	A	*ME*	BD	64658	71492	64701
508 111	**ME**	A	*ME*	BD	64659	71493	64702
508 112	**ME**	A	*ME*	BD	64660	71494	64703
508 114	**ME**	A	*ME*	BD	64662	71496	64705
508 115	**ME**	A	*ME*	BD	64663	71497	64706
508 117	**ME**	A	*ME*	BD	64665	71499	64708
508 120	**ME**	A	*ME*	BD	64668	71502	64711
508 122	**ME**	A	*ME*	BD	64670	71504	64713
508 123	**ME**	A	*ME*	BD	64671	71505	64714
508 124	**ME**	A	*ME*	BD	64672	71506	64715
508 125	**ME**	A	*ME*	BD	64673	71507	64716
508 126	**ME**	A	*ME*	BD	64674	71508	64717
508 127	**ME**	A	*ME*	BD	64675	71509	64718
508 128	**ME**	A	*ME*	BD	64676	71510	64719
508 130	**ME**	A	*ME*	BD	64678	71512	64721
508 131	**ME**	A	*ME*	BD	64679	71513	64722
508 134	**ME**	A	*ME*	BD	64682	71516	64725
508 136	**ME**	A	*ME*	BD	64684	71518	64727
508 137	**ME**	A	*ME*	BD	64685	71519	64728
508 138	**ME**	A	*ME*	BD	64686	71520	64729
508 139	**ME**	A	*ME*	BD	64687	71521	64730
508 140	**ME**	A	*ME*	BD	64688	71522	64731
508 141	**ME**	A	*ME*	BD	64689	71523	64732
508 143	**ME**	A	*ME*	BD	64691	71525	64734

Class 508/2. Units facelifted for the South Eastern lines by Wessex Traincare/ Alstom, Eastleigh 1998–1999.

DMSO. Lot No. 30979 1979–1980. –/66. 36.0 t.
TSO. Lot No. 30980 1979–1980. –/79 1W. 26.5 t.
BDMSO. Lot No. 30981 1979–1980. –/74. 36.5 t.

508 201	(508 101)	**CX**	A	DN	64649	71483	64692
508 202	(508 105)	**CX**	A	DN	64653	71487	64696
508 203	(508 106)	**CN**	A	DN	64654	71488	64697
508 204	(508 107)	**CX**	A	DN	64655	71489	64698
508 205	(508 109)	**CN**	A	DN	64657	71491	64700
508 206	(508 113)	**CX**	A	DN	64661	71495	64704
508 207	(508 116)	**CN**	A	GI	64664	71498	64707
508 208	(508 119)	**CN**	A	DN	64667	71501	64710
508 209	(508 121)	**CX**	A	DN	64669	71503	64712
508 210	(508 129)	**CN**	A	DN	64677	71515	64720
508 211	(508 132)	**CN**	A	DN	64680	71514	64723
508 212	(508 133)	**CX**	A	GI	64681	71511	64724

Class 508/3. Units facelifted units for use on Euston–Watford Junction services by Alstom, Eastleigh 2002–2003.

DMSO. Lot No. 30979 1979–1980. –/68 1W. 36.0 t.
TSO. Lot No. 30980 1979–1980. –/86. 26.5 t.
BDMSO. Lot No. 30981 1979–1980. –/68 1W. 36.5 t.

508 301	(508 102)	**SL**	A	ZG	64650	71484	64693
508 302	(508 135)	**SL**	A	ZG	64683	71517	64726
508 303	(508 142)	**SL**	A	ZG	64690	71524	64733

3. EUROSTAR UNITS (CLASS 373)

Eurostar units were built for and are normally used on services between Britain and Continental Europe via the Channel Tunnel. Apart from such workings units may be used as follows:

SNCF-owned units 3203/04, 3225/26 and 3227/28 have been removed from the Eurostar pool and only operate SNCF-internal services between Paris and Lille. In addition six of the former Regional Eurostar sets are now on hire to SNCF for use on Paris–Lille services and also Paris–Douai–Valenciennes turns.

Each train consists of two 10-car units coupled, with a motor car at each driving end (the sets built for Regional Eurostar services are 8-car). All units are articulated with an extra motor bogie on the coach adjacent to the motor car.

Sets marked "r" have been refurbished. This now includes all sets used by Eurostar, but not 3101/02 (in store) or the sets used by SNCF.

Formation: DM–MSO–4TSO–RB–2TFO–TBFO or DM–MSO–3TSO–RB–TFO–TBFO. Gangwayed within pair of units. Air conditioned.
Construction: Steel.
Supply Systems: 25 kV AC 50 Hz overhead or 3000 V DC overhead (* also equipped for 1500 V DC overhead operation).
Control System: GTO–GTO Inverter on UK 750 V DC and 25 kV AC, GTO Chopper on SNCB 3000 V DC.
Wheel Arrangement: Bo–Bo + Bo–2–2–2–2–2–2–2–2–2.
Length: 22.15 m (DM), 21.85 m (MS & TBF), 18.70 m (other cars).
Couplers: Schaku 10S at outer ends, Schaku 10L at inner end of each DM and outer ends of each sub set.
Maximum Speed: 186 m.p.h. (300 km/h.)
Built: 1992–1993 by GEC-Alsthom/Brush/ANF/De Dietrich/BN Construction/ACEC.
Note: DM vehicles carry the set numbers indicated below.

Class 373/0. 10-Car sets. Built for services starting from/terminating in London Waterloo (now St Pancras). Individual vehicles in each set are allocated numbers 373xxx0 + 373xxx1 + 373xxx2 + 373xxx3 + 373xxx4 + 373xxx5 + 373xxx6 + 373xxx7 + 373xxx8 + 373xxx9, where 3xxx denotes the set number.

Non-standard livery (0): Grey with silver ends, TGV symbol & green/blue doors.

373xxx0 series. DM. Lot No. 31118 1992–1995. 68.5 t.
373xxx1 series. MSO. Lot No. 31119 1992–1995. –/48 2T. 44.6 t.
373xxx2 series. TSO. Lot No. 31120 1992–1995. –/58 1T (r –/56 1T). 28.1 t.
373xxx3 series. TSO. Lot No. 31121 1992–1995. –/58 2T (r –/56 2T). 29.7 t.
373xxx4 series. TSO. Lot No. 31122 1992–1995. –/58 1T (r –/56 1T). 28.3 t.
373xxx5 series. TSO. Lot No. 31123 1992–1995. –/58 2T (r –/56 2T). 29.2 t.
373xxx6 series. RB. Lot No.31124 1992–1995. 31.1 t.
373xxx7 series. TFO. Lot No. 31125 1992–1995. 39/– 1T. 29.6 t.
373xxx8 series. TFO. Lot No. 31126 1992–1995. 39/– 1T. 32.2 t.
373xxx9 series. TBFO. Lot No. 31127 1992–1995. 25/– 1TD. 39.4 t.

3001 r **EU** EU *EU* TI Tread Lightly
3002 r **EU** EU *EU* TI Voyage Vert
3003 r **EU** EU *EU* TI Tri City Athlon 2010
3004 r **EU** EU *EU* TI Tri City Athlon 2010
3005 r **EU** EU *EU* TI
3006 r **EU** EU *EU* TI

Set	Livery	Owner	Pool	Depot	Name
3007 r	**EU**	EU	*EU*	TI	Waterloo Sunset
3008 r	**EU**	EU	*EU*	TI	Waterloo Sunset
3009 r	**EU**	EU	*EU*	TI	REMBERING FROMELLES
3010 r	**EU**	EU	*EU*	TI	REMBERING FROMELLES
3011 r	**EU**	EU	*EU*	TI	
3012 r	**EU**	EU	*EU*	TI	
3013 r	**EU**	EU	*EU*	TI	LONDON 2012
3014 r	**EU**	EU	*EU*	TI	LONDON 2012
3015 r	**EU**	EU	*EU*	TI	
3016 r	**EU**	EU	*EU*	TI	
3017 r	**EU**	EU	*EU*	TI	
3018 r	**EU**	EU	*EU*	TI	
3019 r	**EU**	EU	*EU*	TI	
3020 r	**EU**	EU	*EU*	TI	
3021 r	**EU**	EU	*EU*	TI	
3022 r	**EU**	EU	*EU*	TI	
3101	**EU**	SB		TI	
3102	**EU**	SB		TI	
3103 r	**EU**	SB	*EU*	FF	
3104 r	**EU**	SB	*EU*	FF	
3105 r	**EU**	SB	*EU*	FF	
3106 r	**EU**	SB	*EU*	FF	
3107 r	**EU**	SB	*EU*	FF	
3108 r	**EU**	SB	*EU*	FF	
3201 r*	**EU**	SF	*EU*	LY	
3202 r*	**EU**	SF	*EU*	LY	
3203	**0**	SF	*SF*	LY	
3204	**0**	SF	*SF*	LY	
3205 r	**EU**	SF	*EU*	LY	
3206 r	**EU**	SF	*EU*	LY	
3207 r*	**EU**	SF	*EU*	LY	MICHEL HOLLARD
3208 r*	**EU**	SF	*EU*	LY	MICHEL HOLLARD
3209 r*	**EU**	SF	*EU*	LY	THE DA VINCI CODE
3210 r*	**EU**	SF	*EU*	LY	THE DA VINCI CODE
3211 r	**EU**	SF	*EU*	LY	
3212 r	**EU**	SF	*EU*	LY	
3213 r*	**EU**	SF	*EU*	LY	
3214 r*	**EU**	SF	*EU*	LY	
3215 r*	**EU**	SF	*EU*	LY	
3216 r*	**EU**	SF	*EU*	LY	
3217 r*	**EU**	SF	*EU*	LY	
3218 r*	**EU**	SF	*EU*	LY	
3219 r	**EU**	SF	*EU*	LY	
3220 r	**EU**	SF	*EU*	LY	
3221 r*	**EU**	SF	*EU*	LY	
3222 r*	**EU**	SF	*EU*	LY	
3223 r*	**EU**	SF	*EU*	LY	
3224 r*	**EU**	SF	*EU*	LY	
3225	**0**	SF	*SF*	LY	
3226	**0**	SF	*SF*	LY	
3227	**0**	SF	*SF*	LY	
3228	**0**	SF	*SF*	LY	
3229 r*	**EU**	SF	*EU*	LY	
3230 r*	**EU**	SF	*EU*	LY	
3231 r	**EU**	SF	*EU*	LY	
3232 r	**EU**	SF	*EU*	LY	

Class 373/2. 8-Car sets. Built for Regional Eurostar services, now on long-term hire to SNCF. Individual vehicles in each set are allocated numbers 373xxx0 + 373xxx1 + 373xxx2 + 373xxx3 + 373xxx5 + 373xxx6 + 373xxx7 + 373xxx9, where 3xxx denotes the set number.

3733xx0 series. DM. 68.5 t.
3733xx1 series. MSO. –/48 1T. 44.6 t.
3733xx2 series. TSO. –/58 2T. 28.1 t.
3733xx3 series. TSO. –/58 1T. 29.7 t.
3733xx5 series. TSO. –/58 1T. 29.2 t.
3733xx6 series. RB. 31.1 t.
3733xx7 series. TFO. 39/– 1T. 29.6 t.
3733xx9 series. TBFO. 18/– 1TD. 39.4 t.

Set	Livery	Owner	Pool	Depot	Name
3301	**EU**	EU	*SF*	LY	
3302	**EU**	EU	*SF*	LY	
3303	**EU**	EU	*SF*	LY	
3304	**EU**	EU	*SF*	LY	
3305	**EU**	EU	*SF*	LY	
3306	**EU**	EU	*SF*	LY	
3307	**EU**	EU	*SF*	LY	
3308	**EU**	EU		TI	
3309	**EU**	EU	*SF*	LY	
3310	**EU**	EU	*SF*	LY	
3311	**EU**	EU	*SF*	LY	
3312	**EU**	EU	*SF*	LY	
3313	**EU**	EU	*SF*	LY	ENTENTE CORDIALE
3314	**EU**	EU	*SF*	LY	ENTENTE CORDIALE

Spare DM:

Set	Livery	Owner	Pool	Depot
3999	**EU**	EU	*EU*	TI

4. INTERNAL USE EMUS

The following two vehicles are used for staff training at Virgin's training centre in Crewe. They are from Pendolino 390 033 which was damaged in the Lambrigg accident of February 2007.

390 033	**VT**	VI	Crewe	69133	69833

5. EMUS AWAITING DISPOSAL

The list below comprises vehicles awaiting disposal which are stored on the national railway network.

25 kV AC 50 Hz OVERHEAD UNITS:

390 033	**VT**	VI	LM	69433	69533

Spare car:

Cl. 309	**RR**	WC	CS	71758

750 V DC THIRD RAIL UNITS:

1498	**G**	SW	BM	76773	62411		76844
3905	**CX**	BT	AF	76398	62266	70904	76397
3918	**CX**	BT	AF	76528	62321	70950	76527
930 010	**B/RK**	EM	DY	975600	(10988)	975601	(10843)

Spare cars:

Non-standard liveries:
70293 – Used for paint trials.
76112 – Silver (prototype Class 424 "Networker Classic" conversion).

Cl. 411	**0**	E	ZI	70293
Cl. 424	**0**	BT	ZD	76112

▲ Southern-liveried 377 425 is seen near Reigate with the 13.10 from London Charing Cross on 02/05/09. **Alex Dasi-Sutton**

▼ London Overground-liveried 378 143 arrives at Sydenham with the 15.40 Dalston Junction–West Croydon on 04/06/10. This is one of 20 third-rail only units dedicated to the East London Line. **Robert Pritchard**

▲ One of the new Siemens Class 380s, 4-car set 380 105, stands at Polmadie depot during commissioning on 10/09/10. **Robin Ralston**

▼ Virgin Trains-liveried 390 052 "Virgin Knight" passes Longport with the 17.00 London Euston–Manchester Piccadilly on 15/06/10. **Cliff Beeton**

▲ 29 Hitachi Class 395s are used on the Southeastern High Speed services. On 10/04/10 395 008 arrives at Ebbsfleet International with the 13.44 Dover Priory–London St Pancras. **Antony Guppy**

▼ Gatwick Express-liveried 442 404/401 pass Salfords with the 14.15 London Victoria–Gatwick Airport on 29/05/09. **Brian Denton**

▲ South West Trains white-liveried 444 021/010 pass Basingstoke with the 10.20 Weymouth–Waterloo on 07/02/09. **Brian Denton**

▼ Southern-liveried 455 815 leads 456 007/010 with the 16.52 London Bridge–London Victoria via Crystal Palace at Forest Hill on 04/06/10. **Robert Pritchard**

▲ Southern-liveried 456 010 leads an 8-car 456/455 formation passing Selhurst depot with the 08.50 West Croydon–London Victoria on 27/04/10. **Alex Dasi-Sutton**

▼ A pair of South West Trains white-liveried Class 458s, led by 8019, are seen near Barnes with a Reading–Waterloo service as they overtake 450 544 (one of the Standard Class only 450/5s) on a stopping service on 29/09/09. **Peter Foster**

▲ Gatwick Express-liveried 460 004, with Emirates advertising, is seen near Earlswood with the 14.45 London Victoria–Gatwick Airport on 12/09/09.
Alex Dasi-Sutton

▼ Carrying the new all over white Southeastern livery with blue doors plus a black lower bodyside stripe, 465 025 and 465 180 (still in the older livery with yellow doors) arrive at London Bridge with the 14.49 Plumstead–Cannon Street ecs on 22/09/10. **William Turvill**

▲ Merseyrail-liveried 508 114 arrives at Ellesmere Port with the 09.00 from Liverpool Central on 10/04/10. **Phil Quine**

▼ Eurostar set 3220/19 is seen on High Speed 1 near Harrietsham with the 12.29 London St Pancras–Brussels Midi on 10/10/09. **Chris Wilson**

▲ Blackpool & Fleetwood Tramway Progress Twin Cars 685+675 pass the Queensgate Hotel near Warley Road with a Thornton Gate–Pleasure Beach service on 15/08/10. **Phil Chilton**

▼ A pair of the new Docklands Light Railway cars, 121+105, arrive at Gallions Reach with a Beckton–Tower Gateway train on 06/02/10. **Robert Pritchard**

6. UK LIGHT RAIL & METRO SYSTEMS

This section lists the rolling stock of the various light rail and metro systems in the UK. Passenger carrying vehicles only are covered (not works vehicles). This listing does not cover the London Underground network.

6.1. BLACKPOOL & FLEETWOOD TRAMWAY

Until the opening of Manchester Metrolink, the Blackpool tramway was the only urban/inter-urban tramway system left in Britain. The infrastructure is owned by Blackpool Corporation, and the tramway is operated by Blackpool Transport Services Ltd., using a mixture of trams dating back to the 1930s, as well as some newer vehicles dating from the 1980s. The line normally runs for 11½ miles from Fleetwood in the north to Starr Gate in the south, but is at present undergoing conversion to a modern light rail system for which a new fleet of 16 Bombardier "Flexity 2" trams has been ordered.
System: 550 V DC overhead (600 V DC from 2012).
Depot & Workshops: Rigby Road, Blackpool.
Standard livery: Cream & green except where stated otherwise.

All cars are single-deck unless stated otherwise. For advertising liveries predominating colours are given.

(S) – Stored out of service.

The status of individual trams is flexible and is varied to meet traffic requirements and seasonal demand.

OPEN BOAT CARS A1-1A

Used during the summer season.
Built: 1934 by English Electric. 12 built.
Traction Motors: Two EE327 of 30 kW.
Seats: 56 (* 52).

600
602 * **Yellow & Black**
604 *
605 * (S) **Green & Cream**
607 (S) **Yellow & Green**

Named: 600 "THE DUCHESS OF CORNWALL".

BRUSH CARS A1-1A

Most of the Brush Railcars are stored with only three in use at the time of writing.
Built: 1937 by Brush, Loughborough. 20 built.
Traction Motors: Two EE305 of 40 kW. **Seats**: 48.
Advertising liveries:

621 – Hot Ice Show, Pleasure Beach (blue)
622 – Pontins (blue & yellow)
627 – Buccaneer Family Bar (black)
630 – Karting 2000 (yellow & purple)
631 – Walls ice cream (red)
637 – Blackpool Zoo (green & white)

621 (S)	AL	627 (S)	AL	632			
622 (S)	AL	630	AL	637 (S)	AL		
625 (S)		631	AL				

CENTENARY CLASS A1-1A

The newest trams in use, these are used all year round.
Built: 1984–1987. Body by East Lancs. Coachbuilders, Blackburn. Driver-only operated.
Traction Motors: Two EE305 of 40 kW. **Seats:** 53.
† Rebuilt from GEC car 651.
Advertising liveries:

644 – Farmer Parrs Animal World (yellow)
646 – Paul Gaunt Furniture (blue)
647 – B&M Bargain Stores (black, blue & yellow)

641	**Orange**	643	**Black**	645	**Red**	647	**AL**
642	**Yellow**	644	**AL**	646	**AL**	648 †	**White**

PROGRESS TWIN CARS A1-1A + 2-2

These cars mainly see use during the Illuminations.
Built: Motor cars (671–676) rebuilt 1958–1960 from English Electric Railcoaches by Blackpool Corporation Transport. Driving trailers (681–686) built 1960 by Metro-Cammell.
Traction Motors: Two EE305 of 40 kW. **Seats:** 53 + 61.

671+681	**Green/yellow**	673+683	**Turquoise/yellow**	675+685	**Red/yellow**
672+682	**Orange/yellow**	674+684	**Blue/yellow**	676+686	(S)

ENGLISH ELECTRIC RAILCOACHES A1-1A

Built: Rebuilt 1958–1960 from EE Railcoaches. Originally ran with trailers.
Traction Motors: Two EE305 of 40 kW. **Seats:** 48.
Advertising livery: 678 – Radiowave (black & blue)

678	(S)	**AL**	680	(S)	**Blue**

"BALLOON" DOUBLE DECKERS A1-1A

The "Balloon" cars have been the mainstay of the fleet during the summer months and are also used in lesser numbers during the winter.
Built: 1934–1935 by English Electric. 700–712 were originally built with open tops and 706 has now reverted to that condition.
Traction Motors: Two EE305 of 40 kW. **Seats:** 94 (*† 92, ‡ 90).

Notes: 717 is named "PHILLIP R THORPE"
719 is named "DONNA'S DREAM HOUSE"
* Rebuilt with a new flat front end design and air-conditioned cabs. Known as "Millennium Class".
o 706 – Rebuilt as an open-topped double-decker seating 92. Named "PRINCESS ALICE". Also carries original number 243.

Advertising liveries:

704 – Eclipse at the Globe, Pleasure Beach (black & orange)
707 – Coral Island – The Jewel on the Mile (black)
709 – Blackpool Sealife Centre (blue)
711 – Blackpool Zoo (various)
719 and 721 – Pleasure Beach Resort (black/gold)
723 – Sands Venue nightclub (black)
724 – Lyndene Hotel (blue)
726 – HM Coastguard (blue & yellow)

700		**Green & Cream**	710 (S)	**Yellow/purple**	719		**AL**
701	‡	**Yellow**	711 †	**AL**	720		**White**
704 (S)		**AL**	713	**Yellow/purple**	721		**AL**
706	o		715	**Yellow/blue**	723	†	**AL**
707	*	**AL**	717		724	*	**AL**
708 (S)	‡		718 *	**Yellow/blue**	726		**AL**
709	*	**AL**					

JUBILEE CLASS DOUBLE DECKERS

Built: Rebuilt 1979/1982 from double deckers 725 and 714 respectively. Standard bus ends, Westinghouse Chopper control and stairs at each end. 761 has one door per side whereas 762 has two. Suitable for driver-only operation.
Traction Motors: Two EE305 of 40 kW. **Seats**: 104 (* 86).
Note: 762 is named "STUART L PILLAR"
Advertising liveries:

761 – Wynsors World of Shoes (orange)
762 – www.reblackpool.com (Green/various)

761 **AL**	762 * **AL**

ILLUMINATED CARS

732 (S)	The Rocket	Rebuilt: 1961	Seats: 47
733	Western Train loco & tender	Rebuilt: 1962	Seats: 35
734	Western Train coach	Rebuilt: 1962	Seats: 60
736	"Warship" HMS Blackpool	Rebuilt: 1965	Seats: 71
737	Illuminated Trawler – "Fisherman's Friend"	Rebuilt: 2001	Seats: 48

VINTAGE CARS

These trams are used for special services as well as for occasional normal services, particularly during the Illuminations.

Stockport 5	Open-top double-decker	Built: 1901
Blackpool & Fleetwood 40	Single deck "box car"	Built: 1914
Bolton 66	Bogie double-decker	Built: 1901
Blackpool 143	Open balcony standard double-decker	Built: 1924
Blackpool 147 MICHAEL AIREY	Standard double-decker	Built: 1924
Blackpool 304	Coronation Class single decker	Built: 1952
Sheffield "Roberts Car" 513	Double-decker	Built: 1950
Blackpool 660	Coronation Class single decker	Built: 1953

6.2. DOCKLANDS LIGHT RAILWAY

This system runs for a total of 19 route miles from termini at Bank and Tower Gateway in central London to Lewisham, Stratford, Beckton and Woolwich Arsenal. Another extension from Canning Town to Stratford International is due to open in January 2011. The first line was opened in 1987 from Tower Gateway to Island Gardens. Originally owned by London Transport, it is now part of the London Rail division of Transport for London and operated by Serco Docklands. Cars are normally "driven" automatically using the Alcatel "Seltrack" moving block signalling system.
Notes: Original P86 and P89 Class vehicles 01–21 were withdrawn from service in 1991 (01–11) and 1995 (12–21) and sold for use in Essen, Germany.

55 new cars from Bombardier in Germany entered traffic between 2008 and 2010. These new vehicles have enabled 3-unit trains to start operating.

System: 750 V DC third rail (bottom contact). High-floor.
Depots: Beckton (main depot) and Poplar.
Livery: Red with a curving blue stripe to represent the River Thames.

CLASS B90 2-SECTION UNITS

Built: 1991–1992 by BN Construction, Bruges, Belgium. Chopper control.
Wheel Arrangement: B-2-B. **Traction Motors:** Two Brush of 140 kW.
Seats: 52 (4). **Weight:** 37 t.
Dimensions: 28.80 x 2.65 m. **Braking:** Rheostatic.
Couplers: Scharfenberg. **Maximum Speed:** 50 m.p.h.
Doors: Sliding. End doors for staff use.

22	26	30	34	38	42
23	27	31	35	39	43
24	28	32	36	40	44
25	29	33	37	41	

CLASS B92 2-SECTION UNITS

Built: 1992–1995 by BN Construction, Bruges, Belgium. Chopper control.
Wheel Arrangement: B-2-B. **Traction Motors:** Two Brush of 140 kW.
Seats: 52 (4). **Weight:** 37 t.
Dimensions: 28.80 x 2.65 m. **Braking:** Rheostatic.
Couplers: Scharfenberg. **Maximum Speed:** 50 m.p.h.
Doors: Sliding. End doors for staff use.

45	53	61	69	77	85
46	54	62	70	78	86
47	55	63	71	79	87
48	56	64	72	80	88
49	57	65	73	81	89
50	58	66	74	82	90
51	59	67	75	83	91
52	60	68	76	84	

CLASS B2K 2-SECTION UNITS

Built: 2002–2003 by Bombardier Transportation, Bruges, Belgium.
Wheel Arrangement: B-2-B.
Traction Motors: Two Brush of 140 kW.
Seats: 52 (4).
Weight: 37 t.
Dimensions: 28.80 x 2.65 m.
Braking: Rheostatic.
Couplers: Scharfenberg.
Maximum Speed: 50 m.p.h.
Doors: Sliding. End doors for staff use.

92	96	01	05	09	13
93	97	02	06	10	14
94	98	03	07	11	15
95	99	04	08	12	16

CLASS B07 2-SECTION UNITS

Built: 2007–2010 by Bombardier Transportation, Bautzen, Germany.
Wheel Arrangement: B-2-B.
Traction Motors: Two Brush of 140 kW.
Seats: 52 (4).
Weight: 37 t.
Dimensions:
Braking: Rheostatic.
Couplers: Scharfenberg.
Maximum Speed: 50 m.p.h.
Doors: Sliding. End doors for staff use.

101	111	120	129	138	147
102	112	121	130	139	148
103	113	122	131	140	149
104	114	123	132	141	150
105	115	124	133	142	151
106	116	125	134	143	152
107	117	126	135	144	153
108	118	127	136	145	154
109	119	128	137	146	155
110					

6.3. EDINBURGH TRAMWAY

A new tramway is under construction in Edinburgh, with the first of the new trams now complete. The route runs for 11½ miles from Newhaven to Edinburgh Airport via Leith and central Edinburgh, including the famous Princes Street, and Waverley and Haymarket stations. The scheme has been delayed by construction problems but it is currently expected that at least an initial section will open in 2012. The trams will be the longest to operate in the UK.

System: 750 V DC overhead.
Platform Height: 350 mm.
Depot & Workshops: Gogar.
Livery: White, red & black.

CAF — 7-SECTION TRAMS

Built: 2009–2011 by CAF, Irun, Spain.
Wheel Arrangement: Bo-Bo-2-Bo.
Seats: 78.
Dimensions: 42.8 x 2.65 m.
Couplers:
Doors: Sliding plug.
Traction Motors: 12 CAF of 80 kW.
Weight:
Braking:
Maximum Speed: 50 m.p.h.

251	257	263	268	273
252	258	264	269	274
253	259	265	270	275
254	260	266	271	276
255	261	267	272	277
256	262			

6.4. GLASGOW SUBWAY

This circular 4 ft. gauge underground line is the smallest metro system in the UK, running for just over six miles. It is commonly referred to as the "Subway" or the "Clockwork Orange". Operated by Strathclyde PTE the system has 15 stations. The entire passenger railway is underground, contained in twin tunnels, allowing for clockwise operation on the "outer" circle and anti-clockwise operation on the "inner" circle.

Trains are formed of 3-cars – either three power cars or two power cars sandwiching one of the newer trailer cars.

System: 600 V DC third rail.
Depot & Workshops: Broomloan.
Livery: Strathclyde PTE carmine & cream unless stated.

SINGLE POWER CARS

Built: 1977–1979 by Metro-Cammell, Birmingham. Refurbished 1993–1995 by ABB Derby.
Wheel Arrangement: Bo-Bo.
Traction Motors: Four GEC G312AZ of 35.6 kW each.
Seats: 36.
Couplers: Wedglock.
Weight: 19.6 t.
Dimensions: 12.81 m x 2.34 m.
Doors: Sliding.
Maximum Speed: 33.5 m.p.h.

101	108	115	122	128
102	109	116	123	129
103	110	117	124	130
104	111	118	125	131
105	112	119	126	132
106	113	120	127	133
107	114	121		

INTERMEDIATE BOGIE TRAILERS

Built: 1992 by Hunslet Barclay, Kilmarnock.
Seats: 40.
Couplers: Wedglock.
Weight: 17.2 t.
Dimensions: 12.70 m x 2.34 m.
Doors: Sliding.
Maximum Speed: 33.5 m.p.h.
Advertising/Special liveries:

201 – Scottish Sun (white/red/black).
203 – Radio Clyde1 (red).
204 – SPT Zonecard ticket (blue).
205 – Robert Burns (different images of the famous poet).

201	**AL**	203	**AL**	205	**AL**	207	208
202		204	**AL**	206			

6.5. GREATER MANCHESTER METROLINK

Metrolink was the first modern tramway system in the UK, combining street running with longer distance running over former BR lines. The system opened in 1992 from Bury to Altrincham with a street section through the centre of Manchester and a spur to Piccadilly station. A second line opened in 2000 from Cornbrook to Eccles extending the total route mileage to 23 miles. A short spur off the Eccles line to MediaCityUK opened in September 2010. Further extensions are under construction as follows:

- Rochdale station via Oldham to (involving converting the former National Rail line to light rail use) between spring 2011 and spring 2012.
- The East Manchester Line to Droylsden (spring 2012) and Ashton-under-Lyne (winter 2013–14).
- The South Manchester Line to St Werburgh's Road (spring 2011), East Didsbury (summer 2013) and Manchester Airport (mid 2016).

Further extensions will see trams running into Oldham and Rochdale town centres (2014) and a second city crossing in Manchester.

Operator: Stagecoach Metrolink.
System: 750 V DC overhead. High floor.
Depot & Workshops: Queens Road, Manchester. A second depot is under construction at Trafford to service the expanding fleet.

1000 SERIES 2-SECTION TRAMS

Built: 1991–1992 by Firema, Italy. Chopper control.
Wheel Arrangement: Bo-2-Bo. **Traction Motors**: Four GEC of 130 kW.
Dimensions: 29.0 x 2.65 m. **Seats**: 82 (4).
Doors: Sliding. **Couplers**: Scharfenberg.
Weight: 45 t. **Maximum Speed**: 50 m.p.h.
Braking: Rheostatic, regenerative, disc and emergency track.

Livery: White, dark grey & blue with light blue doors.

* Fitted with front-end valances, retractable couplers and controllable magnetic track brakes for on-street running mixed with private vehicles.

No.		Name	No.		Name
1001	*		1014		THE GREAT MANCHESTER RUNNER
1002	*	DA VINCI	1015	*	BURMA STAR
1003	*		1016	*	
1004			1017		BURY HOSPICE
1005	*		1018		
1006	*		1019		
1007	*	EAST LANCASHIRE RAILWAY	1020	*	LANCASHIRE FUSILIER
1008			1021	*	
1009	*		1022	*	POPPY APPEAL
1010	*		1023		
1011	*		1024		
1012	*		1025	*	
1013			1026		

2000 SERIES 2-SECTION TRAMS

Built: 1999 by Ansaldo, Italy. Chopper control. Fitted with front-end valances, retractable couplers and controllable magnetic track brakes for on-street running mixed with private vehicles.
Wheel Arrangement: Bo-2-Bo.
Traction Motors: Four GEC of 130 kW.
Dimensions: 29.0 x 2.65 m.
Seats: 82 (4).
Doors: Sliding.
Couplers: Scharfenberg.
Weight: 45 t.
Maximum Speed: 50 m.p.h.
Braking: Rheostatic, regenerative, disc and magnetic track.

Livery: White, dark grey & blue with light blue doors.

2001 (S)
2002
2003
2004
2005
2006

3000 SERIES FLEXITY SWIFT 2-SECTION TRAMS

A total of 62 new Bombardier B5000 "Flexity Swift" trams have been ordered to strengthen services on existing routes and for the extensions listed above.
Built: 2009–2012 by Bombardier, Vienna, Austria.
Wheel Arrangement: Bo-2-Bo.
Traction Motors: Four Bombardier 3-phase asynchronous of 120 kW.
Dimensions: 28.4 x 2.65 m.
Seats: 52.
Doors: Sliding.
Couplers: Scharfenberg.
Weight: 39.7 t.
Maximum Speed: 50 m.p.h.
Braking: Rheostatic, regenerative, disc and magnetic track.

Livery: New Manchester Metrolink silver & yellow.

3001
3002
3003
3004
3005
3006
3007
3008
3009
3010
3011
3012
3013
3014
3015
3016
3017
3018
3019
3020
3021
3022
3023
3024
3025
3026
3027
3028
3029
3030
3031
3032
3033
3034
3035
3036
3037
3038
3039
3040
3041
3042
3043
3044
3045
3046
3047
3048
3049
3050
3051
3052
3053
3054
3055
3056
3057
3058
3059
3060
3061
3062

6.6. LONDON TRAMLINK

This system runs through central Croydon via a one-way loop, with lines radiating out to Wimbledon, New Addington and Beckenham Junction/Elmers End, the total route mileage being 18½ miles. It opened in 2000 and is now operated by Transport for London.

System: 750 V DC overhead. **Platform Height**: 350 mm.
Depot & Workshops: Therapia Lane, Croydon.

BOMBARDIER 3-SECTION TRAMS

Built: 1998–1999 by Bombardier-Wien Schienenfahrzeuge, Austria.
Wheel Arrangement: Bo-2-Bo. **Traction Motors**: Four of 120 kW each.
Dimensions: 0.1 x 2.65 m. **Seats**: 70.
Doors: Sliding plug. **Couplers**: Scharfenberg.
Weight: 36.3 t. **Maximum Speed**: 50 m.p.h.
Braking: Disc, regenerative and magnetic track.

Livery: Light grey & lime green with a blue solebar.

2530	2534	2538	2542	2546	2550
2531	2535	2539	2543	2547	2551
2532	2536	2540	2544	2548	2552
2533	2537	2541	2545	2549	2553

Name: 2535 STEPHEN PARASCANDOLO 1980–2007

6.7. NOTTINGHAM EXPRESS TRANSIT

This was the last light rail system in the UK, opened in 2004. Line 1 runs for 8¾ miles from Station Street, Nottingham (alongside Nottingham station) to Hucknall, including a short spur to Phoenix Park. There is around three miles of street running through Nottingham. Extensions are planned to Clifton (Line 2) to the south of Nottingham, and Chilwell via Beeston to the west (Line 3).

The system is operated by the Arrow Light Rail Ltd. consortium (Transdev, Nottingham City Transport, Carillion, Bombardier, Innsfree and Galaxy).

System: 750 V DC overhead. **Platform Height**: 350 mm.
Depot & Workshops: Wilkinson Street.

BOMBARDIER INCENTRO 5-SECTION TRAMS

Built: 2002–2003 by Bombardier, Derby Litchurch Lane Works.
Wheel Arrangement: Bo-2-Bo. **Traction Motors**: 8 x 45 kW wheelmotors.
Dimensions: 33.0 x 2.4 m **Seats**: 54 (4).
Doors: Sliding plug. **Couplers**: Not equipped.
Weight: 36.7 t. **Maximum Speed**: 50 m.p.h.
Braking: Disc, regenerative and magnetic track for emergency use.

Standard livery: Black, silver & green unless stated.
Advertising liveries:

201 – Nottinghamcontemporary.org (yellow/light blue & white).
211 – NET "tram It" (various).

201	**AL**	Torvill and Dean
202		DH Lawrence
203		Bendigo Thompson
204		Erica Beardsmore
205		Lord Byron
206		Angela Alcock
207		Mavis Worthington
208		Dinah Minton
209		Sid Standard
210		Sir Jesse Boot
211	**AL**	Robin Hood
212		William Booth
213		Mary Potter
214		Dennis McCarthy
215		Brian Clough

6.8. MIDLAND METRO

This system opened in 1999 and has one 12½ mile line from Birmingham Snow Hill to Wolverhampton along the former GWR line to Wolverhampton Low Level. On the approach to Wolverhampton it deviates from the former railway alignment to run on-street to the St. George's terminus. It is operated by Travel West Midlands Ltd. Extensions are proposed from Snow Hill through Birmingham to Five Ways and from Wednesbury to Brierley Hill and Dudley.

System: 750 V DC overhead.
Platform Height: 350 mm.
Depot & Workshops: Wednesbury.

ANSALDO 2-SECTION TRAMS

Built: 1998–1999 by Ansaldo Transporti, Italy.
Wheel Arrangement: Bo-2-Bo.
Traction Motors: Four x 105 kW each.
Dimensions: 24.00 x 2.65 m.
Seats: 52 (4).
Doors: Sliding plug.
Couplers: Not equipped.
Weight: 35.6 t.
Maximum Speed: 43 m.p.h.
Braking: Rheostatic, regenerative, disc and magnetic track.

Standard livery: Dark blue & light grey with green stripe, yellow doors & red front end.

MW: New Network West Midlands tram livery (silver & pink).
Note: 01 is stored out of use and used for spares.

01	(S)	SIR FRANK WHITTLE
02		
03		RAY LEWIS
04		
05	**MW**	SISTER DORA
06		ALAN GARNER
07	**MW**	BILLY WRIGHT
08		JOSEPH CHAMBERLAIN
09	**MW**	JEFF ASTLE
10	**MW**	JOHN STANLEY WEBB
11		THERESA STEWART
12		
13		ANTHONY NOLAN
14		JIM EAMES
15		AGENORIA
16		GERWYN JOHN

6.9. SHEFFIELD SUPERTRAM

This system opened in 1994 and has three lines radiating from Sheffield City Centre. These run to Halfway in the south-east, with a spur from Gleadless Townend to Herdings Park, to Middlewood in the north with a spur from Hillsborough to Malin Bridge and to Meadowhall Interchange in the north east, adjacent to the large shopping complex. The total route mileage is 18 miles. The system is a mixture of on-street and segregated running.

The cars are owned by South Yorkshire Light Rail Ltd., a subsidiary of South Yorkshire PTE. The operating company, South Yorkshire Supertram Ltd. is leased to Stagecoach who operate the system as Stagecoach Supertram.

Because of severe gradients in Sheffield (up to 1 in 10) all axles are powered on the vehicles, which have low-floor outer sections.

System: 750 V DC overhead. **Platform Height**: 450 mm.
Depot & Workshops: Nunnery.
Standard livery: Stagecoach (All over blue with red & orange ends).

Advertising livery: 120 – East Midlands Trains (blue).

SIEMENS 3-SECTION TRAMS

Built: 1993–1994 by Siemens, Krefeld, Germany.
Wheel Arrangement: B-B-B-B.
Traction Motors: Four monomotor drives of 250 kW.
Dimensions: 34.75 x 2.65 m. **Seats**: 80 (6).
Doors: Sliding plug. **Couplers**: Not equipped.
Weight: 52 t. **Maximum Speed**: 50 m.p.h.
Braking: Rheostatic, regenerative, disc and emergency track.

101	106	110	114	118	122
102	107	111	115	119	123
103	108	112	116	120 **AL**	124
104	109	113	117	121	125
105					

6.10. TYNE & WEAR METRO

The Tyne & Wear Metro system covers 48 route miles and can be described as the UK's first modern light rail system.

The initial network opened between 1980 and 1984 consisting of a line from South Shields via Gateshead and Newcastle Central station to Bank Foot (later extended to Newcastle Airport in 1991) and the North Tyneside loop (over former BR lines) serving North Shields, Tynemouth and Whitley Bay with a terminus at St. James in Newcastle city centre. A more recent extension came from Pelaw to Sunderland and South Hylton in 2002, making use of existing heavy rail infrastructure between Heworth and Sunderland.

The system is owned by Nexus – the Tyne & Wear PTE and operated by DB Regio.
System: 1500 V DC overhead. **Depot & Workshops**: South Gosforth.

METRO-CAMMELL 2-SECTION UNITS

Built: 1978–1981 by Metropolitan Cammell, Birmingham (Prototype cars 4001 and 4002 were built by Metropolitan Cammell in 1976 and rebuilt 1984–1987 by Hunslet TPL, Leeds).
Wheel Arrangement: B-2-B.
Traction Motors: Two Siemens of 187 kW each.
Dimensions: 27.80 x 2.65 m.
Seats: 68.
Doors: Sliding plug.
Couplers: BSI.
Weight: 39.0 t.
Maximum Speed: 50 m.p.h.

Standard livery: Red & yellow unless otherwise indicated.
B Blue & yellow
G Green & yellow.
O (4001) Original 1975 Tyne & Wear Metro livery of yellow & cream.
O (4027) Original North Eastern Railway style (red & white).
Advertising liveries:

4002 – Tyne & Wear Metro (orange & black).
4020 – Modern Apprenticeships (white, red & blue).
4038 – Talktofrank.com (white).
4040 – Cut your CO_2 day (blue & white).
4042 – Metro Radio (blue & pink).
4045 – Newcastle Racecourse (green).
4049 – Kidd & Spoor Harper Solicitors (blue).
4055 – European Regional Development Fund (blue & yellow).
4075 – Tyne & Wear Public Services (purple & white).
4080 – South Shields market (white).

4001	**O**	4019		4037		4055	**AL**	4073	
4002	**AL**	4020	**AL**	4038	**AL**	4056		4074	
4003		4021		4039	**B**	4057		4075	**AL**
4004	**G**	4022		4040	**AL**	4058	**B**	4076	**B**
4005		4023	**G**	4041		4059		4077	
4006		4024	**B**	4042	**AL**	4060		4078	
4007		4025	**G**	4043		4061		4079	
4008		4026		4044		4062	**G**	4080	**AL**
4009		4027	**O**	4045	**AL**	4063		4081	**B**
4010		4028		4046		4064		4082	
4011		4029	**B**	4047	**B**	4065		4083	**B**
4012		4030		4048		4066	**B**	4084	
4013		4031	**B**	4049	**AL**	4067		4085	
4014		4032		4050		4068		4086	
4015		4033		4051	**G**	4069		4087	
4016	**B**	4034		4052		4070		4088	
4017		4035	**B**	4053	**B**	4071		4089	
4018	**G**	4036	**G**	4054	**B**	4072	**B**	4090	

Names:

4026	George Stephenson
4041	HARRY COWANS
4060	Thomas Bewick
4064	Michael Campbell
4065	DAME Catherine Cookson
4073	Danny Marshall
4077	Robert Stephenson
4078	Ellen Wilkinson

7. CODES

7.1. LIVERY CODES

Code Description

1 "One" (metallic grey with a broad black bodyside stripe. White National Express "interim" stripe as branding).
AL Advertising/promotional livery (see class heading for details).
B BR blue.
C2 c2c Rail (blue with metallic grey doors & pink c2c branding).
CN Connex/Southeastern (white with black window surrounds & grey lower band).
CX Connex (white with yellow lower body & blue solebar).
EU Eurostar (white with dark blue & yellow stripes).
FB First Group dark blue.
FS First Group (indigo blue with pink & white stripes).
FU First Group "Urban Lights" (varying blue with pink, white & blue markings on the lower bodyside).
G BR Southern Region or BR green.
GE First Great Eastern (grey, green, blue & white).
GV Gatwick Express EMU (red, white & indigo blue with mauve & blue doors).
HC Heathrow Connect (grey with a broad deep blue bodyside band & orange doors).
HE Heathrow Express (grey & indigo blue with black window surrounds).
LM London Midland (grey & green with broad black stripe around the windows).
LO London Overground (all over white with a blue solebar & black window surrounds).
LT London Transport maroon & cream.
ME Merseyrail (metallic silver with yellow doors).
N BR Network SouthEast (white & blue with red lower bodyside stripe, grey solebar & cab ends).
NC National Express white (white with blue doors).
NO Northern (deep blue, purple & white).
NX National Express (white with grey ends).
0 Non-standard livery (see class heading for details).
RK Railtrack (green & blue).
RM Royal Mail (red with yellow stripes above solebar).
RR Regional Railways (dark blue/grey with light blue & white stripes, three narrow dark blue stripes at cab ends).
SB Southeastern High Speed (all over blue with black window surrounds).
SC Strathclyde PTE (carmine & cream lined out in black & gold).
SD South West Trains outer suburban livery {Class 450 style} (deep blue with red doors & orange & red cab sides).
SE Southeastern (all over white with light blue doors and (on some units) dark blue lower bodyside stripe).
SL Silverlink (indigo blue with a white stripe, green lower body & yellow doors).
SN Southern (white & dark green with light green semi-circles at one end of each vehicle. Light grey band at solebar level).

SP Strathclyde PTE {revised} (carmine & cream, with a turquoise stripe).
SR ScotRail – Scotland's Railways (dark blue with Scottish Saltire flag & white/light blue flashes).
SS South West Trains inner suburban {Class 455} (red with blue & orange flashes at unit ends).
ST Stagecoach {long-distance stock} (white & dark blue with dark blue window surrounds and red & orange swishes at unit ends).
TL New Thameslink (silver with blue window surrounds and ends).
VT Virgin Trains silver (silver, with black window surrounds, white cantrail stripe & red roof. Red swept down at unit ends. Black & white striped doors).
WN Old West Anglia Great Northern (white with blue, grey & orange stripes).
YR West Yorkshire PTE/Northern EMUs (red, lilac & grey).

7.2. OWNER CODES

A Angel Trains
BT Bombardier Transportation UK
E Eversholt Rail (UK)
EM East Midlands Trains
EU Eurostar (UK)
HE British Airports Authority
LY Lloyds TSB
P Porterbrook Leasing Company
QW QW Rail Leasing
RM Royal Mail
SB SNCB/NMBS (Société Nationale des Chemins de fer Belges/Nationale Maatschappij der Belgische Spoorwegen)
SF SNCF (Société Nationale des Chemins de fer Français)
SW South West Trains
VI Virgin Trains
WC West Coast Railway Company

7.3. OPERATOR CODES

C2 c2c
DB DB Schenker
EA National Express East Anglia
EU Eurostar (UK)
FC First Capital Connect
HC Heathrow Connect
HE Heathrow Express
LM London Midland
LO London Overground
ME Merseyrail
NO Northern
SE Southeastern
SF SNCF (French Railways)
SN Southern
SR ScotRail
SW South West Trains

7.4. ALLOCATION & LOCATION CODES

Code	*Location*	*Depot Operator*
AD	Ashford	Hitachi
AF	Ashford Chart Leacon Works	Bombardier Transportation
BD	Birkenhead North (Liverpool)	Merseyrail
BF	Bedford Cauldwell Walk	First Capital Connect
BI	Brighton Lovers Walk	Southern
BM	Bournemouth	South West Trains
CE	Crewe International	DB Schenker
CS	Carnforth	West Coast Railway Company
DN*	Donnington Railfreight Terminal (Telford)	*Storage location only*
DY	Derby Etches Park	East Midlands Trains
EM	East Ham (London)	c2c
FF	Forest (Brussels)	SNCB/NMBS
GI	Gillingham (Kent)	Southeastern
GW	Glasgow Shields Road	ScotRail
HE	Hornsey (London)	First Capital Connect
IL	Ilford (London)	National Express East Anglia
LG	Longsight (Manchester)	Northern
LM	Long Marston (Warwickshire)	Motorail Logistics
LY	Le Landy (Paris)	SNCF
MA	Manchester Longsight	Alstom
NG	New Cross Gate (London)	London Overground
NL	Neville Hill (Leeds)	East Midlands Trains/Northern
NN	Northampton King's Heath	Siemens
NT	Northam (Southampton)	Siemens
OH	Old Oak Common Heathrow (London)	Heathrow Express
RM	Ramsgate	Southeastern
RY	Ryde (Isle of Wight)	South West Trains
SG	Slade Green (London)	Southeastern
SL	Stewarts Lane (London)	Southern/VSOE
SO	Soho (Birmingham)	London Midland
SU	Selhurst (Croydon)	Southern
TI	Temple Mills (London)	Eurostar
WB	Wembley (London)	Alstom
WD	Wimbledon (London)	South West Trains
ZA	RTC Business Park (Derby)	Railway Vehicle Engineering
ZB	Doncaster Works	Wabtec Rail
ZC	Crewe Works	Bombardier Transportation
ZD	Derby Works	Bombardier Transportation
ZG	Eastleigh Works	Knights Rail Services
ZH	Springburn Depot (Glasgow)	Railcare
ZI	Ilford Works	Bombardier Transportation
ZJ	Stoke Works	Axiom Rail (Stoke)
ZK	Kilmarnock Works	Brush-Barclay
ZN	Wolverton Works	Railcare
ZR	York (Holgate Works)	Network Rail

*= unofficial code.